The Outstanding Universal Value and Conservation of Hubei Shennongjia

Zongqiang Xie • Guozhen Shen

The Outstanding Universal Value and Conservation of Hubei Shennongjia

The World Natural Heritage Site

Springer

Science Press
Beijing

Zongqiang Xie
State Key Laboratory of Vegetation and Environmental Change
Institute of Botany, CAS
Beijing, China

Guozhen Shen
State Key Laboratory of Vegetation and Environmental Change
Institute of Botany, CAS
Beijing, China

ISBN 978-981-16-0683-0 ISBN 978-981-16-0684-7 (eBook)
https://doi.org/10.1007/978-981-16-0684-7

Jointly published with Science Press

This Springer imprint is published by the registered company Springer Nature Singapore Pte Ltd.
The registered company address is: 152 Beach Road, #21-01/04 Gateway East, Singapore 189721, Singapore

Other Contributors

Dayong Fan, Xianming Gao, Jielin Ge, Chunqing Li, Liyan Li, Daxing Wang, Zhixian Wang, Gaoming Xiong, Wenting Xu, Changming Zhao, and Youbing Zhou.

Preface

Designation of protected areas is a key measure for safeguarding species and ecosystems globally. In response to massive worldwide biodiversity loss, the global extent of protected land has roughly doubled in size, with more than 202,000 protected areas now covering 14.7% of the world's terrestrial area. By the end of 2018, China, one of the world's megadiverse countries, had established 2750 nature reserves, encompassing 147 million ha and covering 15.3% of China's land area. Natural World Heritage, recognized globally as its outstanding universal value, had become the pinnacle of natural protected areas that are the cornerstones of biodiversity conservation. Until 2020, totally 249 protected areas, including 38 cultural and natural mixed heritage, were inscribed on the World Heritage List, among which 18 heritage sites were admitted from China.

The mountains of Central China was the linkage corridors between the flora of Sino-Japan and the flora of Sino-Himalaya, and the joint zone of South-west, East China, and Taiwan flora. Hubei Shennongjia, located in the center of Central China, maintained the function and stability of the ecosystems in Qinba Mountain and north subtropical Mountain due to its rich diversity and typical biological and ecological process. Since the beginning of 19th century, researchers had conducted lots of investigations and studies on geological landscape, biodiversity, and ecosystem functions of Hubei Shennongjia. All of the works had lay foundations for uncovering the outstanding universal value of Hubei Shennongjia.

Here, we compiled literature and conducted additional field surveys in Hubei Shennongjia to illustrate the outstanding universal value of Hubei Shennongjia according to World Heritage criteria (ix) and (x), following the operational guidelines for the implementation of the World Heritage Convention. Our study provides great insight into protecting, monitoring, and managing the most outstanding elements and its integrity of the heritage in the Northern Hemisphere.

Acknowledgment

We wish to acknowledge Junsheng Ying, Keping Ma, Xianming Gao, Dayong Fan, Gaoming Xiong, Wenting Xu, Youbing Zhou, Jielin Ge, Yanjun Du, Jiangshan Lai, and Yan Zhu from Institute of Botany, the Chinese Academy of Sciences (IBCAS). We thank Zhigang Jiang from Institute of Zoology, Chinese Academy of Sciences (IOZCAS), Xiaoping Zuo and Hongchun Liu from Ministry of Housing and Urban-Rural Development of the People's Republic of China. We thank all the field investigators and supports by Liyan Li, Daxing Wang, Chunqing Li, and Zhixian Wang from Administrative Bureau of Shennongjia Nature Reserve. We also thank Rong Tian from CCTV Science and Education Program Production Center and Na Wang from Phoenix Satellite Television for providing image data. We are grateful to the editors from Science Press for their organizational coordination during the editing and publication of the book.

This work has been financially supported by the State Key Research and Development Program of the Chinese Ministry of Science and Technology (No. 2019YFD1100403), and National Observation and Research Station for Forest Ecosystem in Shennongjia, Hubei, China.

In spite of our best efforts, errors still remain in the book due to our limited knowledge and information on Shennongjia. Please send us your comments so that we can update the book in the near future.

Contents

Chapter 1
Natural Condition

Shennongjia is located on the eastern edge of the second level of terrain in China (China's terrain has a three-level, ladder-like distribution from west to east, with the first level being the elevated Qinghai–Tibetan plateau, the second level being the mountainous central areas, and the third being the low lying eastern plains). Shennongjia is also in a transition zone, where the subtropical climate changes to a warm temperate climate characterized by the north subtropical monsoon climate. The landform of Shennongjia belongs to the subalpine mountains of the Eastern Daba mountain range. The area's unique geographical location and climatic characteristics make it an area of both transition and intersection for plants and animals and mean it is rich in species. As a result, Shennongjia preserved the most intact forest vegetation of all the areas located in the north subtropical zone (latitude of N 30°). This preservation means that it is honored as "The Green Miracle" in the Tropic of Cancer. Shennongjia plays a significant role as one of the key regions for studying and conserving biodiversity, as well as one of the three distribution centers of seed plant genera which are endemic to China.

1.1 Location

Shennongjia is located in the northern part of Hubei Province adjacent to Chongqing Municipality. The area stretches east to west from Baokang County, Hubei, to Wuxi County, Chongqing, extending 140 km from E110°3′10″ to E110°34′4″. It stretches north to south from Zhuxi County, Hubei, to Badong County, Hubei, covering 144 km from N31°35′49″ to N31°21′21″.The main part of Shennongjia is situated in the Shennongjia Forest District of Hubei Province (Fig. 1.1). However, as a mountain range, Shennongjia also extends to about ten other counties including the Hubei counties of Xingshan, Badong, Zigui, Fangxian, Zhushan, Zhuxi, Baokang, and the Chongqing counties of Wushan and Wuxi (Xu et al. 2019).

Z. Xie, G. Shen, *The Outstanding Universal Value and Conservation of Hubei Shennongjia*, https://doi.org/10.1007/978-981-16-0684-7_1

Fig. 1.1 Location of the Hubei Shennongjia in China

1.2 Geological Features

Hubei Shennongjia is located in the Shennongjia fault dome, which is the watershed of the Yangtze and Han Rivers. The vault keeps the horizontal orientation of the Sinian System and its back is wide and flat. The stratum in the north flank is flat, with a dip angle of less than 20°. Meanwhile on the south flank the stratum is much steeper, with an angle of more than 20°. The south edge is a group of diagonal side curtain folds that were formed between the Cambrian and the Triassic periods. The nearly east-west distributed stratum of the Hubei Shennongjia is exposed

Photo 1.1 Dawokeng Formation

completely, especially those from the Pre-Cambrian and Lower Paleozoic periods. They are mainly exposed on Da Shennongjia (Big Shennongjia Peak), Xiao Shennongjia (Little Shennongjia Peak), and Laojunshan (Laojun Mountain), etc. The stratum is mostly comprised of slight, regional, metamorphic rocks: dolomite, sandstone, conglomerate, slate, phyllite and basaltic rocks. Outcropped strata are mostly igneous rocks which can be divided into two categories, intrusive rocks and volcanic rocks, according to their characteristics. Intrusive rocks consist of diabase, gabbro, peridotite, distributing at the top of Shennongding Mountain, as well as Laojunshan (Laojun Mountain) and Huangbaoping. Volcanic rocks are mainly basalt, with minerals dominantly consisting of slender cylindrical plagioclase, pyroxene, and basic glass. These rocks are mainly distributed in the Shennongjia Luanshigou Formation, the Dawokeng Formation, the Taizi Formation, and the Macaoyuan Group (Photos 1.1 and 1.2).

1.3 Topography

Shennongjia sits on the eastern edge of the area forming the second level of China's west to east, three-level, ladder-like distribution of terrain. The area has mountain landforms that are the eastern part of the Daba Mountains. Shennongjia's east-west mountain range is consistent with the regional geographical structure, being higher in the southwest and lower in the northeast, and featuring some tall and majestic mountains and deep canyons. The majority of the peaks are above 1500 m above sea

Photo 1.2 Shicaohe Formation

Photo 1.3 Landform of Yinyu River in Hubei Shennongjia

level. More specifically, there are 46 peaks higher than 2500 m above sea level, and six peaks higher than 3000 m above sea level. Among them, Da Shennongjia (Big Shennongjia) has an elevation of 3052 m and Xiao Shennongjia (Little Shennongjia) has an elevation of 3005 m (Photos 1.3 and 1.4). The highest peak is Shennongding

Photo 1.4 Chief peak in Central China—Shennong Peak

(at 3106.2 meters above sea level), which is also the highest point in the Daba Mountains and in Central China. The peak is also known as "the Roof of Central China." The minimum altitude is 400 m at Shizhu River (Xiaguping Township) with a relative elevation difference of 2700 m.

1.4 Climate

Shennongjia belongs to the north subtropical monsoon climate zone, where the weather is warm and humid. Therefore, the area's climate is significantly different from the dry climate in the desert influenced by the subtropical anticyclone at the same latitude. Shennongjia is in an area of transition between China's subtropical zone and its warm temperate zone. Cold and warm air masses from the north and the south, respectively, meet here, controlled by the subtropical anticyclone. Shennongjia has four distinct seasons and has a favorable amount of rainfall and temperature range. Due to its undulating terrain, Shennongjia has a vertical climatic spectrum from its mountains' bases to their peaks. This spectrum can be divided into north subtropical belt, warm temperate belt, temperate belt, and cold temperate belt. The mean annual temperature is 12.1 °C. The average temperature in the coldest month (January) is minus 8 °C and is 26.5 °C in the hottest month (July). The mean annual sunshine is 1858.3 h and the frost-free period lasts 217 days (Photos 1.5 and 1.6).

Photo 1.5 Snow cover in winter—Liangfengya

Photo 1.6 Summer Scenery—Luoyanghe valley

The precipitation in Shennongjia shows obvious seasonality and accounts for 86.8% of the annual total precipitation from April to October. The spatial distribution diversity of precipitation leads to an overall trend of reducing from south to north and increasing from the foot to the peak. The mean annual evaporation is 500–800 mm, and the drought index is from 0.50 to 0.53 (Zhu & Song, 1999).

1.5 Hydrology

Hubei Shennongjia has a vigorous water system which distributes radially around the Da and Xiao Shennongjia (Big and Small Shennongjia Peaks). There are four major river systems, including the Xiangxi River, the Yandu River, the Nan River, and the Du River. The Hubei Shennongjia has a total river area of 16.39 km^2 and its total annual amount of runoff is 2.2 billion m^3 (Photo 1.7).

1.6 Soil

Shennongjia is located in the transition area between China's subtropical and warm temperate zones. In terms of landforms, it lies between the eastern plains and the hills and western mountains and plateaus. Therefore, diverse hydrothermal conditions produce remarkable differences in the vertical structure of the soil on the area's different slopes. Moreover, Shennongjia has apparent altitudinal soil belts. The belt spectrum of soil of the south hillside is as follows: At 1000 m and below 1000 m, it is

Photo 1.7 Source of Xiangxi river

mountain yellow soil of broad-leaved evergreen forests; From 1000 m to 1700 m, it is mountain yellow brown soil of mixed evergreen and deciduous broad-leaved forests; From 1700 m to 2200 m, it is mountain brown soil of deciduous broad-leaved forests; From 2200 m to 2600 m, there are coniferous and broad-leaved mixed forests and these forests have mountain dark brown soil; Finally, above 2600 m, mountain podzolized dark brown soil is found in meadow and farges fir forest. The altitudinal soil spectrum of the north hillside is different: Below 800 m, there is yellow brown soil of broad-leaved evergreen forests; From 800 m to 1600 m, there is mountain yellow brown soil of mixed evergreen and deciduous broad-leaved forests (Photo 1.8); From 1600 m to 2100 m there is mountain brown soil of deciduous

Photo 1.8 Typical soil in Hubei Shennongjia—Mountain yellow brown soil

broad-leaved forests; From 2100 m to 2500 m there is mountain dark brown soil of coniferous and broad-leaved mixed forests; From 2500 m to 3000 m there is podzolized dark brown soil of mountain meadow and farges fir forests (Zhu & Song, 1999).

Chapter 2
History

2.1 Natural History

Hubei Shennongjia is located on the northern margin of the Yangtze plate. Its rock stratum contains all the units since the Mesoproterozoic era. In the Mesoproterozoic era (a period from about 1600–1000 million years ago), crustal extension resulted in a continental margin rift basin in the northern margin of the Yangtze plate. This led to the formation of the Proterozoic Shennongjia Group, which mainly contains dolomite and is rich in stromatolites. It has a sedimentary thickness of 4000 meters. This is one of the most comprehensive outcropping geological units of the Yangtze plate. It is also the best preserved late-Pre-Cambrian stratigraphic unit in the world.

During a period from 1000 to 800 million years ago, the Greenville orogenic movement (the Jinnin orogenic event) swept the world. This incident made the world a unified super Rodinia continent by joining several secondary continents together. Afterwards, in the Hubei Shennongjia, the ocean turned into land, the sea basin became mountains. Since then, the world has entered into a stage of continental splitting. With the expansion of the continental crust, deep magmatism occurred, forming a ring-shaped, basic dyke. During the same period, due to the "Snowball Earth" event, a thick glacial deposit accumulated in the opened basin, namely the Nantuo Tillite of the Nanhua System. The glacial deposit is significant evidence that records the ancient climate and geographic location of Shennongjia at that time.

In the early Paleozoic era (from 540 to 408 million years ago), the Hubei Shennongjia underwent transgression and regression alternately several times. In the late Paleozoic era (from 408 to 251 million years ago), it was a marine environment and then it became land till the middle Devonian period. Primordial Lepidodendrales and Prolycopods plants dominated, while some Psilopsida (such as Psilophytites) and primitive Filicopsida (such as Archaeopterids) plants also grew. In the late Devonian period, the oldest Archaeopteris suborder plants had their golden age, while the primitive Sphenophyta plants also flourished. In the Mesozoic era (a period between about 251 and 65 million years ago), the geological

Z. Xie, G. Shen, *The Outstanding Universal Value and Conservation of Hubei Shennongjia*, https://doi.org/10.1007/978-981-16-0684-7_2

movements of the Hubei Shennongjia fell into a pattern similar to that of modern times. In this period, the Indo-China and Yanshan tectonic movements occurred and along with them came a block uplift of the stratum, which laid the basic outline for the dome-shaped landscape of mountains in Shennongjia. In the early Triassic period, cycads, conifers, and ginkgo began to emerge here because of the warm and humid climate.

In the Cenozoic era (from 65 million years ago to the present), the area, affected by the Himalayan orogeny, has been continuing to uplift in fits and starts, forming Shennongjia, the highest peak in Central China. The area's climate changed to one which was hot and dry, which has helped the gymnosperm and angiosperm plants grow lush. By the time the cold weather of the Tertiary period came to the area, a large number of angiosperms had appeared here. Since the start of the Quaternary period (from 2.5 million years ago to the present), with a background of an overall uplift of Shennongjia, Neotectonics have performed some intermittent arch rises and new fracture events, which have given Shennongjia geomorphic features such as high mountains and deep valleys. In addition, because of the towering mountains in the area, the area continued to develop during the Quaternary Glaciation. Whilst glaciers and a cold climate were widespread around the world, the glaciers had little effect on the vegetation of Shennongjia (Zhu & Song, 1999). The vegetation quickly restored because of its location at a lower altitude and the area's complex and varied terrain conditions. As a result, some rare and ancient higher plants (such as *Tetracentron sinense*, *Euptelea pleiosperma*, *Parakmeria lotungensis*, etc.) now still survive here.

In summary, a large number of ancient and relic species have been retained in the Hubei Shennongjia and many of them are remnants from the Paleogene and Cretaceous periods. Therefore, the property plays a significant role in studies of the world's biota, as well as studies of the biological evolution. Thus, Hubei Shennongjia is of global scientific significance.

2.2 Human History

Hubei Shennongjia has a long history. Archeologists have excavated stone implements made and used about 1.2 million years ago in the Chaoyang river valley. In Hongping, they have also discovered an ancient cave called Xiniudong which was a site of human activity about 100,000 years ago. Much evidence indicates that Hubei Shennongjia had an excellent ecological environment and rich biological resources which meant it provided an ideal living environment for early humans. In the pre-Qin period (before 202 BC), Hubei Shennongjia was part of the Chu Kingdom for 800 years. In the Qin and Han dynasties (from 202 BC to 220 AD), Fangling County was settled and governed by the people of Hanzhong County in Shaanxi province. However, after the Han Dynasty (from 220 AD to 1643 AD), Hubei Shennongjia was governed by its neighboring state and county. Later on, in the Qing Dynasty (from 1643 AD to 1911 AD), it belonged to the Fang County affiliated to Yunyang Prefecture and Xingshan County affiliated to Yichang Prefecture in Hubei Province.

From ancient times, Hubei Shennongjia served as a land to which deposed kings were sent to live in exile. From the pre-Qin period to the Song Dynasty, 14 kings were exiled to Fang County. Among these kings, the most famous one was Li Xian, an emperor of the Tang Dynasty, who was banished here in the 686 AD and stayed in exile for 13 years. The exiled kings had great influences on Shennongjia, creating a lot of beautiful legends and giving deep cultural heritage to Shennongjia.

Hubei Shennongjia was a refuge during the transition periods between different Chinese dynasties. Hubei Shennongjia has always been regarded as a place where few people tread because of its high mountains, dense forests, thick fog, steep cliffs, and deep valleys. Whenever dynasties changed and infighting between local warlords led to crushing poverty, then Hubei Shennongjia was the first choice for people to flee to in order to avoid the disasters. Hubei Shennongjia is also on the route of the Ancient Salt Road. The Ancient Salt Road in Shennongjia was known as the "Southern Silk Road," which crossed the Yangtze-Han Rivers and Sichuan-Hubei. The Ancient Salt Road connected with Jing (Hubei) and Xiang (Hunan) in the east and connected to Shi (Enshi) and Yi (Yichang) in the south. It connected to Ba (Chongqing) and Shu (Sichuan) in the west. It was the main road for the transportation of Sichuan's salt to the Central Plains and had a history lasting about 1500 years (begin at 618 AD). During these years, a unique Salt Merchants Culture was formed. Hubei Shennongjia boasts rich human resources, beautiful and ancient legends, as well as ancient and mysterious folk customs. There is a widespread legend that a Chinese ancestor named Shen Nong tasted a hundred herbs and pioneered Chinese agricultural civilization. There are also many myths related to the area and other intangible cultural heritage. Along with the fact that the Ancient Salt Road crossed Sichuan and Hubei for more than 1000 years, there are relics of ancient troops and folk customs in the area.

2.3 History of Protection

Hubei Shennongjia is located in the deep mountains of the northwest of Hubei Province. The area was not widely known because of its steep terrain, vegetation coverage, and difficult transportation. Consequently, the outstanding resources of Hubei Shennongjia were not recognized by scholars until the mid-1950s. Therefore, its resources and environment have been kept intact. Because of its precipitous landforms, primitive vegetation cover, and difficult transportation, the population density in Shennongjia is low. This relatively closed environment also provides a natural barrier to effectively protect biodiversity.

2.3.1 *Human Activities*

2.3.1.1 Population

The total area of Hubei Shennongjia is 73,318 ha, with 6999 permanent residents. The density of population is 10.4 people per square kilometer. This is less than 25 people per square kilometer, and so the area is considered as a sparsely populated area. The population is mainly distributed at the north-west part of Hubei Shennongjia including Dongxi village, Huangboqian village, and the south part of Hubei Shennongjia including Pingqian village, Qingshu village, Banqiao village, Xinglongsi village, and Jinjiaping village. One-third of local residents go out to work, one-third are engaged in forest management and protection and paid by government, one-third are engaged in tea industry, beekeeping, and other agricultural activities. The total area of the buffer zone is 41,536 ha, with 7388 permanent residents. The density of population is 17.8 people per square kilometer, so it also a sparsely populated area. Permanent residents in the buffer zone are mainly distributed at Muyu Town. They are mainly engaged in forest tea industry, beekeeping, tourism service, and other activities.

2.3.1.2 Major Human Activities

Forest Management and Protection

In the past, deforestation has occurred in Hubei Shennongjia. After the establishment of the nature reserve, deforestation was comprehensively forbidden and vegetation recovery was conducted using different measures at different elevations and in different types of topography (Wang et al., 2017).

Tourism Activities

Hubei Shennongjia has beautiful scenery and rich tourist resources. Eco-tourism in Hubei Shennongjia began in the late 1980s. Currently this eco-tourism has reached a considerable scale. The annual number of tourists exceeds 522,000. Tourism activities are mainly concentrated in the peak season from July to October (Yu et al., 2018b). The management authorities of Hubei Shennongjia strictly control the construction of local infrastructure, tour activities, and the number of tourists. They employ scientific management and scientific decision-making. A series of appropriate rules have also been developed to protect the ecological environment of Hubei Shennongjia. There are many regulations and ordinances protecting the natural and ecological environment of Hubei Shennongjia. As a result, tourism has had little effect on the natural and ecological environment.

Land Use Pattern

More than 90% of the land in the heritage and its buffer zone is woodland. The main land use types by human are dry land, orchard (tea industry), urban land, and rural residential land, and are mainly distributed in the buffer zone. There are very few lands for scenic facilities in Hubei Shennongjia. There were some dry lands in the northern part (Qinshulin, Dongxi, and Luoyanghe), the southern part (Xiangsiling, Banqiao, and Xiaguping), the eastern part (Laojunshan) of Hubei Shennongjia, they have been abandoned and are being transformed to woodlands after the projects of "Protect Primeval Forest," "Return Farmland to Forest," and "The Yangtze River Protection Forest Engineering."

Scientific Research

There are many precious ancient records that discuss Hubei Shennongjia, such as the *Compendium of Materia Medica*. Many famous people have also vividly described Hubei Shennongjia. The book *Xici* first described Shennongjia and the Shennong Clan. Then the book *Shen Nong's Herbal Classics,* compiled by Tao Hongjing in the Jin dynasty, recorded 365 kinds of traditional Chinese medicines, giving their names, flavors, the illnesses they treated, and their growing environment. From the mid-nineteenth century to the early twentieth century, Russia, the United States, Britain, Germany, France, Sweden, Japan, and other countries all organized expeditions (in the name of governments or individuals) to collect plants and conduct scientific activities in Shennongjia. Among them, the British naturalist Ernest Henry Wilson visited west Hubei four times between 1888 and 1910. He published the book titled with *Natural Scientist in West China* and *China-Mother of Gardens.* These described the characteristics of rare plants in Shennongjia in detail. Wilson also collected many seeds from Shennongjia and bred many plants which were then cultivated throughout Europe. Later a book titled *Plantae Wilsonianae* was compiled by Charles Sprague Sargent, the director of the Arnold Arboretum of Harvard University. The book was a milestone in the early studies of Shennongjia. In addition, the Swedish botanist Harry Smith explored Shennongjia during the period from 1922 to 1934. In 1868, a French biologist named Armand David visited Shennongjia and later published *Plantae Davidianae.* Between 1884 and 1886, a Russia geographer Gregory Nikolayevich Patanin studied in Shennongjia and published the article *Plantae Chinensis Potaninianeneonon Piasekianae.*

Since its inception in the early twentieth century, Chinese scientists have collaborated to study the geology, topography, flora, fauna, and climate of Shennongjia. The Institute of Botany, the Chinese Academy of Sciences have systematically inspected the flora and fauna of Shennongjia five times, between 1922 and 1925, then between 1941 and 1943, then between 1946 and 1947, from 1976 to 1978, and again between 2002 and 2006. Following this research, they have produced books such as *Shennongjia Exploratory Report, Shennongjia Forest Survey Report, Vegetation and Flora of Western Hubei Shennongjia Region, Shennongjia Plants,*

Scientific Expedition Set of Shennongjia Nature Reserve, Rare and Endangered Wild Fauna and Flora Atlas of Shennongjia National Nature Reserve (Zhu & Song, 1999). During this period, some scientific research organizations also did researches on the geology, geography, soil and biology of Shennongjia. Up to now, more than 620 articles and papers on the geology and geomorphology, physical geography, biology and ecology of Hubei Shennongjia have been published by researchers from China and abroad. The area has become an important base for education, research, and field studies for many universities and research institutes in China.

2.3.2 Government Protection and Management

The state and local government now pay great attention to the protection of the environment and resources in the property. Shennongjia has retained two stone tablets which were engraved in the years 1862 and 1887. These bear the inscriptions "forbidden forest" and "forbidden stone and woods." These two stone tablets protected the Shennongjia forests for a long time. During the period of the Republic of China, the government organized two large-scale exploratory activities in Shennongjia forests. These activities resulted in the publication of two reports on about the Shennongjia forests resources—*Shennongjia Exploratory Report (1943)* and *Shennongjia Forest Survey Report (1947)*. These two reports provide important information to support the conservation and management of Shennongjia.

In 1982, the Shennongjia nature reserve was established by Government of Hubei Province. Then in 1986 the State Council approved the establishment of a national forest and wildlife nature reserve. In 1990, the Hubei Shennongjia joined the UNESCO World Network of Biosphere Reserves. Then in 1995 it became one of the ten first nature reserves in China funded by the Global Environment Facility Trust Fund (GEF). In 2006, the reserve was included in the first group of demonstration protected areas of the national forest system. Then in 2009, the Administrative Bureau of Shennongjia National Nature Reserve drew up a master plan which aimed to strengthen the management and protection of the area and effectively protect its ecological environment and biological diversity (Wang et al., 2017). In 2011, Shennongjia became a national forest tourism demonstration area and a national Level 5A tourist attraction. The "Shennongjia Scenic Area Management Regulations" were formulated in order to expand Shennongjia's protection and management. In 2013, the area became a world geopark approved by UNESCO.

2.3.3 Protection of Local Pledges

The local residents of Hubei Shennongjia, including minority, had a tradition of developing regulations to protect the ecological environment and natural resources since ancient times. This tradition still plays an important role today. The local regulations included closing hillsides to facilitate afforestation, strictly forbidding digging, quarrying, deforestation, building kilns, hunting, shooting, and fishing. Such agreements have become widely-acknowledged civil regulations and rules that are worked out and abided by the residents jointly.

2.3.4 Natural Conservation Traditions of Local Inhabitants

Local indigenous residents, including people of the Tujia, Miao, and Dong ethnic minorities, have a culture and religious beliefs that include respect for nature and they think that all things in nature are holy. To many of them, the sacred mountains and old woods are holy and inviolable. One should be punished if one violates these holy things. In addition, many people not only do not kill animals but also have the custom of releasing them, because they are Buddhists. "The more trees and grass you plant, the fewer natural disasters will happen; the more flowers and trees, the happier life you will lead." Awareness about environmental protection is rooted in the local nomadic people's mindsets. Local peoples have a tradition of treasuring nature and protecting the environment in order to ensure their survival.

Chapter 3
Plant Diversity

The complex terrain, the dramatic altitudinal difference, the various soil types, and the climate conditions have all helped Hubei Shennongjia form diverse habitats and incubate abundant biodiversity. There are 11 types of vegetation, 46 formations, and intact altitudinal vegetation spectrums from low to high altitude within the 2700 m elevation range, including evergreen broad-leaved forest, mixed evergreen and deciduous broad-leaved forest, deciduous broad-leaved forest, mixed coniferous and broad-leaved forest, coniferous forest, and bush/meadow (Editorial Committee of Vegetation Map of China, Chinese Academy of Sciences, 2007). Shennongjia has an integral ecosystem and the largest primary forests of central China (Ban & Qi, 2001).

After several geological changes, the geology and geomorphology of Shennongjia developed into its unique geographical environment. This unique environment is one where a lot of rare animals and plants have survived up to now. As such, the area provides real evidences for the evolutionary history of ancient organisms. In addition, it is located in the interface of Sino-Japan Flora and Sino-Himalaya Flora. It is an area where different flora can contact, mix together, and specialize (Xie et al., 2017). As a result, Hubei Shennongjia is a hotspot of wildlife diversity with obvious transitional characteristics (Xie & Shen, 2018). There are nearly 3509 vascular plant species, accounting for approximately 12.5% of the vascular plants in China (Gao et al., 2019). As a hub for temperate flora to evolve and expand, Shennongjia also has the most densely distributed primitive temperate genera and relic genera (Xie et al., 2017). The fauna of Shennongjia mainly belongs to the Oriental region-India-China subkingdom -Central China region-West mountain plateau subdistrict. The fauna exhibits evidence of blend species between the Oriental realm and the Palaearctic realm (Zhou & Wu, 2019). The distribution pattern of animals here is dominated by Oriental and South China types, with some types of Palaearctic, Himalaya-Hengduan Mountains, Northeast and North. Fauna is dominated by the Oriental realm, while also have some characteristics of the Palaearctic realm. These features demonstrate a primitive fauna of abundant

Z. Xie, G. Shen, *The Outstanding Universal Value and Conservation of Hubei Shennongjia*, https://doi.org/10.1007/978-981-16-0684-7_3

biodiversity. There are more than 600 vertebrate species and 4300 insects in Shennongjia region (Xie & Shen, 2018).

Hubei Shennongjia, located at the eastern boundary of the second step of the Chinese terrain, is in the transition region where the land shifts from the eastern plains and hills to the western mountain plateau (Xu et al., 2019). It is also an area of climate transition, with the climate changing from the subtropical zone to the warm temperate zone. The unique geographical position and climate characteristics make the property an area that has a combination of different types of flora and fauna and breeds extensive biodiversity. It makes it an area that has offered a refuge for wild animals and plants since the Quaternary era glaciers and has provided special habitats for lots of ancient relic, rare, endangered, and endemic species.

3.1 Flora

Hubei Shennongjia is situated within the scope of the eastern Asian flora and the joint zone of the flora of Sino-Japan and the flora of Sino-Himalaya. During the area's long geological history, the natural surroundings have experienced several great changes that have provided opportunities for different flora to contact each other, mix together, and specialize. This has led to plentiful biodiversity in Hubei Shennongjia, which has 268 families, 1206 genera, and 3767 higher plant species (including subspecies, varieties, and forms) (Gao et al., 2019). Hubei Shennongjia is also the centralized distribution area for a large number of primeval temperate genera and abundant relic plants. Hubei Shennongjia is a central zone for the origination, differentiation, and development of temperate flora (Table 3.1).

The flora of Hubei Shennongjia, largely consisting of temperate plants, is not only an important part of the East Asian flora, but also has a close phylogenetic relationship with the flora of North America. The seed plant genera in China have been classified into 15 distribution types by Wu Zhengyi (Wu, 2010). According to Wu's classification, Hubei Shennongjia possesses 15 distribution types. Hubei Shennongjia has 77 genera which belong to the cosmopolitan type, counting for 7.7% of the all 1000 genera of wild seed plants. There are 323 genera belonging to the tropic distribution types, or 32.3% of the all these genera. There are 590 genera of the temperate distribution types, or 59% of all these (Gao et al., 2019).

Table 3.1 Statistics for the wild high plants present in Hubei Shennongjia

Category	Families	%	Genera	%	Species	%
Bryophyta	56	20.9	119	9.9	258	6.8
Pteridophyta	39	14.6	87	7.2	325	8.6
Gymnospermae	6	2.2	16	1.3	33	0.9
Angiospermae	167	62.3	984	81.6	3151	83.6
Total	268	100.0	1206	100.0	3767	99.9

Photo 3.1 Chinese katsuratree (*Cercidiphyllum japonicum*), an example of Sino-Japanese distribution subtype and the ancient relic species

The flora of Hubei Shennongjia is mostly of temperate distribution types, especially the north temperate ones. Hubei Shennongjia contains 147 genera of the East Asia distribution type. Among these 147 genera, there are 55 genera belonging to the Sino-Japanese subtype, 36 belonging to the Sino-Himalayan subtype, and 56 widely distributed in areas from the Himalayas to Japan. This shows that Hubei Shennongjia is located in the core of the area where the Sino-Himalayan flora and Sino-Japanese flora intersect each other (Gao et al., 2019) (Photo 3.1). Among Hubei Shennongjia's flora, a certain proportion is also of the East Asia and North America disjuncted type including 81 genera. This flora shows the historical origin and phylogenetic relationship between East Asia and North America. At the same time, the presence of this unique distribution type also shows that the flora in Hubei Shennongjia has ancient significance.

3.2 Ancient and Relic Plants

The beneficial natural conditions present in Hubei Shennongjia have been able to conserve a huge number of ancient plants that originated in the Tertiary period. There are 3767 species of wild higher plants, belonging to 1189 genera and 268 families. Among these plants, 139 families including 597 genera originated

Photo 3.2 Tetracentron (*Tetracentron sinense*), an ancient relic species in Hubei Shennongjia

during the Tertiary period, accounting for 51.9% of the total families and 49.5% of the total genera of such plants (Gao et al., 2019) (Photo 3.2).

The ancient geological data has shown that Hubei Shennongjia emerged from the sea during the Devonian period. Plant fossils of original Lepidodendrales, Lycopsida, and Psilopsida discovered in the late Devonian stratum within Hubei Shennongjia indicate that the first terrestrial plants originated in Shennongjia as early as the middle Devonian period. Plant fossils have been dated back to the late Triassic period, including 11 genera of ferns, seven genera of seed ferns, 11 genera of Cycads, five genera of Ginkgos, and two genera of Conifers. This clearly indicates that vascular plants had been luxuriant in Hubei Shennongjia during the late Mesozoic era. In the stratum of the Cretaceous and the Tertiary periods, a lot of plant and sporopollen fossils, including species from about 60 families, have been found. The fossils are Taxodiaceae, Ephedraceae, Salicaceae,

Photo 3.3 Spikemoss (*Selaginella* sp.)

Photo 3.4 Farges fir (*Abies fargesii*)

Myricaceae, Juglandaceae, Betulaceae, Fagaceae, Ulmaceae, Proteaceae, Santalaceae, Chenopodiaceae, Magnoliaceae, and Hamamelidaceae (Gao et al., 2019).

The fossil evidence has proven that there are a lot of relic species in the ancient plants in Hubei Shennongjia. The plants that were present during these ancient periods of geologic history were mainly genera of *Lycopodium*, *Selaginella* (Photo 3.3), *Osmunda*, *Dicranopteris*, *Lygodium*, *Pinus*, *Juglans*, *Ulmus*, *Taxus*, *Fagus*, *Magnolia* in the Paleozoic Era, and *Abies* (Photo 3.4), *Cyclocarya*, *Salix*, *Liriodendron* in the third century. This fossil evidence has indicated that the distribution area of these ancient plants founded in Shennongjia has shrunk

Photo 3.5 Dove Tree (*Davidia involucrata*)

considerably. For example, the Chinese dove tree (Photo 3.5) nowadays can be found in Hubei Shennongjia and the provinces of Yunnan, Sichuan, Guizhou, as well as further eastward in Hunan and Hubei, and on Mount Huangshan in Anhui and Tianmu Mountain in Zhejiang. However, its sporopollen fossils have been found in Qingjiang County, Jiangxi Province, which is far away from these areas of its modern distribution. The fossil of *Tapiscia* is found in the stratum of the Miocene Epoch in Shandong Province's Shanwang town (Linqu County, Shandong Province). *Fortunearia* possesses only one species, and now it distributes in Hubei Shennongjia and Shaanxi, Henan, Sichuan, Hubei, Anhui, Jiangxi, and Zhejiang Provinces, but its fossils were found in the late Pliocene stratum in Japan. Discovery of fossils has proven that the endemic genera are ancient. The evidence of fossils can be combined with the disjuncted distribution patterns of flora in Hubei Shennongjia. Together, this makes it clear that many ancient plants in Hubei Shennongjia are also relic. *Heptacodium* and *Triaenophora* are typical examples. There is only one species of *Heptacodium* (*H. miconioides*) which is discontinuously distributed in Shennongjia, Hubei Province and in eastern Zhejiang Province. *Triaenophora* is an oligotypic genus of the two species (Photo 3.6), and one living in Jianshi and Badong Counties, Hubei Province, where Hubei Shennongjia is located, and the other living in Chengkou County, Chonqing Municipality. Another example is a similar species, *Spirostegia bucharica*, from the monotypic genus *Spirostegia*, which distributes in Uzbekistan and Turkmenistan in Central Asia. *Triaenophora* lives in steep cliffs of limestone areas, but *Spirostegia* comes from Middle Asia and grows in higher salinity desert areas.

Another two oligotypic genera, *Lirodendron* and *Sassafras*, belong to the East Asia and North America disjuncted distribution type. The former has two species disjunctedly distributed in China (*Liriodendron chinense*) (Photo 3.7), and North America (*L. tulipifera*). The latter possesses three species disjunctedly distributed in mainland China (*Sassafras tzumu*), Taiwan (*S. randaiense*), and North America

Photo 3.6 Rockycabbage (*Triaenophora rupestris*)

(*S. albidum*) (Gao et al., 2019). These species are not only proof of the genetic relationship between the floras of East Asia and North America, but are also typical representatives of relic plants.

All the facts mentioned above show that the flora of Hubei Shennongjia originated as early as the Tertiary period. They show that the towering and complex topography of Shennongjia was a refuge for important living things during the Quaternary period's glaciations when the climate kept fluctuating. As such, a lot of plants in Hubei Shennongjia originated before the Quaternary glaciations and among them there are many ancient and relic species. The habitats of Hubei Shennongjia have provided advantaged conditions for species living, reproducing, and developing.

Photo 3.7 Chinese tuliptree (*Liriodendron chinense*)

3.3 Endemic Plants

Specific geographical position, diversity of topographic features, as well as the specificity of its climate, have made Shennongjia rich in endemic plants (Table 3.2). Totally, there are 205 species and two genera endemic to Hubei Shennongjia, and 1793 species endemic to China (Table 3.3) (Gao et al., 2019; Xie et al., 2017).

3.3.1 Plants Endemic to Shennongjia

Hubei Shennongjia has 205 species which are endemic to the Shennongjia region. These species are classified into 57 families and 131 genera. They include two genera, *Zhengyia* (Urticaceae) and *Repenticaulia* (Apiaceae), recently found and named. In the components of families and genera of the 205 species which are endemic to Shennongjia, the largest portion (or 25 species in 11 genera) is part of the composite family (Gao et al., 2019; Xie et al., 2017). Then 12 species in eight genera are part of the rose family and 11 species in eight genera are part of the Labiatae family. Most of the species in a single genus are found in *Impatiens* (Balsaminaceae), *Parasenecio* (Asteraceae), and *Corydalis* (Papaveraceae), all of which possess seven species endemic to Shennongjia. The remaining genera usually only possess 1–5 species.

According to the life forms of these endemic species, there are 14 trees and 48 shrubs. Then there are 40 deciduous and 22 evergreen woody plants, while the

Table 3.2 Richness of wildlife in Hubei Shennongjia compared with neighboring areas

	Vascular plants		Amphibians		Reptiles		Birds		Mammals	
Type	Species	%[a]	Species	%[a]	Species	%[a]	Species	%[a]	Species	%[a]
Shennongjia	3509		37		53		399		87	
Three gorges area	5081	69.06	43	86.05	44	120.45	415	96.14	99	87.88
Hubei	3928	89.33	69	53.62	78	67.95	521	76.58	121	71.90
Chongqing	5928	59.19	57	64.91	63	84.13	417	95.68	117	74.36
Shaanxi	3569	98.32			53	100.00	465	85.81	147	59.18
China	31,142	11.27	410	9.02	403	13.15	1271	31.39	609	14.29

[a]Species in Hubei Shennongjia as % of species in the area being compared

Table 3.3 The plant species endemic to Shennongjia

Category	Endemic to Shennongjia	Endemic to China	Rate (%)	Note
Pteridophyta	9	86	10.5	
Gymnospermae	1	31	3.2	Endemic species only
Angiospermae	195	1676	11.6	Endemic species only
Total	205	1793	11.4	

Photo 3.8 Zhengyia (*Zhengyia shennongensis*), a new genus and species of Urticaceae

Photo 3.9 Henry figwort (*Scrophularia henryi*)

Photo 3.10 Hongping Apricot (*Armeniaca hongpingensis*)

Photo 3.11 Snapweed (*Impatiens compta*)

rest are herbages (Photos 3.8, 3.9, 3.10, 3.11, 3.12, 3.13, 3.14, 3.15, 3.16, 3.17) (Gao et al., 2019; Xie et al., 2017).

3.3.2 Plants Endemic to China

Within Hubei Shennongjia, there are 1793 species endemic to China, accounting for 51.1% of the vascular plants of Shennongjia. These 1793 plants are classified into

Photo 3.12 Chinese chirita (*Chirita tenuituba*)

Photo 3.13 Shennongjia wormwood (*Artemisia shennongjiaensis*)

136 families and 540 genera, including Pteridophyta (18 families, 37 genera, and 86 species), Gymnospermae (6 families, 15 genera, and 31 species) and Angiospermae (112 families, 488 genera, and 1676 species) (Photos 3.18 and 3.19) (Gao et al., 2019; Xie et al., 2017).

Photo 3.14 Shenlungchia clematis (*Clematis shenlungchiaensis*)

Photo 3.15 Shennongjia monkshood (*Aconitum shennongjiaense*)

3.4 Rare and Endangered Plants

Totally, 234 species have been classified as rare and endangered wild plants in Hubei Shennongjia property (Table 3.4) (Photos 3.20 and 3.21). Among these plants, 110 species are listed on the IUCN Red List (2014). Among which, seven species

Photo 3.16 Shennongjia sandwort (*Arenaria shennongjiaensis*)

Photo 3.17 North magnoliavine (*Schisandra incarnate*)

are termed Critically Endangered (CR), 33 species are Endangered (EN), and 70 species are Vulnerable (VU) (Qin et al., 2017). Of Shennongjia's rare and endangered species, 94 have been listed in the Convention on International Trade in Endangered Species of Wild Fauna and Flora (CITES) (2014) (Table 3.4). These species include 16 species in Appendix I and 78 species in Appendix II. Apart from two varieties (namely *Taxus wallichiana* var. *chinensis* and *T. wallichiana* var. *mairei*), all of the other 92 species listed in CITES belong to the Orchidaceae family, including those such as *Changnienia amoena*, *Cypripedium japonicum*, and *Gastrodia elata*. In addition, 167 of the species found in Shennongjia are on the list of national protected plants, among which 20 species marked as Class I, and

Photo 3.18 Scape monkshood (*Aconitum scaposum*), endemic to China

Photo 3.19 Chinese maple (*Acer sinense*)

147 species marked as Class II (Gao et al., 2019; Xie et al., 2017). All the species in the Orchidaceae family are included (Photos 3.22, 3.23, 3.24, 3.25, 3.26). Rare and endangered species in Hubei Shennongjia are mainly distributed in damp valleys and some special areas characterized by little human disturbances.

Table 3.4 Rare and endangered plants in Hubei Shennongjia

Category	IUCN red list (2014)			CITES (2014)		List of national protected plants	
	CR	EN	VU	Appendix I	Appendix II	Class I	Class II
Pteridophyta	0	2	4	0	0	0	1
Gymnospermae	0	0	5	2	0	2	8
Angiospermae	7	31	61	14	78	18	138
Total	7	33	70	16	78	20	147

Photo 3.20 Chinese Yew (*Taxus wallichiana* var. *chinensis*)

Photo 3.21 Manyleaf paris (*Paris polyphylla*)

Photo 3.22 Chinese umbrella leaf (*Diphylleia sinensis*)

Photo 3.23 Uniflower orchid (*Changnienia amoena*)

There are six concentrated distribution areas including Jiuchong River basin, the key habitat for Chinese yew, dove tree, patung manglietia (*Manglietia patungensis*), martin michelia (*Michelia martini*), orchid plants such as dendrobium, kiwi plants such as coriaceousleaf actinidia (*Actinidia rubricaulis* var. *coriacea*); Yangquan River basin, the key habitat for Chinese katsuratree, farges torreya (*Torreya fargesii*), Chinese yew, orchid plants such as Japanese lady's slippers (*Cypripedium*

Photo 3.24 Henry emmenopterys (*Emmenopterys henryi*)

Photo 3.25 Pashan paris (*Paris bashanensis*)

japonicum) and common pleione (*Pleione bulbocodioides*), kiwi plants such as yangtao actinidia (*Actinidia chinensis*) and kolomikta-vine actinidia (*Actinidia kolomikta*); Yinyu River basin, the key habitat for dove tree, Chinese katsuratree, Chinese yew, manyleaf paris (*Paris polyphylla*), kiwi plants such as Henry actinidia

Photo 3.26 Common pleione (*Pleione bulbocodioides*)

(*Actinidia callosa* var. *henryi*) and fourcalx actinidia (*Actinidia tetramera*), orchid plants such as twoflower rattlesnake plantain (*Goodyera biflora*); Changping River basin, the key habitat for Chinese yew, farges torreya, Chinese katsuratree, serrate clubmoss (*Huperzia serrata*), orchid plants such as franchet lady's slippers (*Cypripedium franchetii*), kiwi plants such as fourcalx actinidia (*Actinidia chinensis* var. *deliciosa*); Majia River basin, the key habitat for Taiwan beech (*Fagus hayatae*), dove tree, Chinese yew, orchid plants such as yellowflower bletilla (*Bletilla ochracea*), kiwi plants such as patung actinidia (*Actinidia tetramera* var. *badongensis*); Shashuwan area, the key habitat for dove tree, Chinese yew, farges clethra (*Clethra fargesii*), kiwi plants such as yangtao actinidia, orchid plants such as common calanthe (*Calanthe discolor*).

3.5 Important Type Specimen Plant Hotspot

Totally, 523 type specimen plants have been collected from the Hubei Pan-Shennongjia region, including Shennongjia Forest District as well as Badong, Xingshan, Baokang, Fang, Zhushan, Zhuxi, and Yichang Counties in Hubei Province and Wushan, Wuxi in Chongqing Municipality (Xie & Xiong, 2020). These 523 species account for 15.3% of all the species of vascular plants in Hubei Shennongjia (Photos 3.27, 3.28, 3.29, 3.30, 3.31, 3.32, 3.33, Table 3.5) (Shan et al., 2019; Xie et al., 2017).

Photo 3.27 Wushan epimedium (*Epimedium wushanense*)

Photo 3.28 Undulata stranvaesia (*Stranvaesia davidiana* var. *undulata*)

Photo 3.29 Denseflower loosestrife (*Lysimachia congestiflora*)

Photo 3.30 Shinningleaf beech (*Fagus lucida*)

Photo 3.31 Henry anise tree (*Illicium henryi*)

Photo 3.32 Henry viburnum (*Viburnum henryi*)

Photo 3.33 Franchet maple (*Acer franchetii*)

Table 3.5 The type specimen plants in Hubei Pan-Shennongjia region

Category	Families	Genera	Species
Pteridophyta	8	14	22
Gymnospermae	2	4	5
Angiospermae	78	245	496
Total	88	263	523

Chapter 4
Animal Diversity

4.1 Fauna

Hubei Shennongjia is located in the hinterland of East Asia, which is the first transition zone from the lowland plains of eastern Eurasia to the mountains of central Eurasia. It also lies in the transition region from the Oriental Realm to the Palaearctic Realm. In addition, it is an area of climactic transition from north subtropical to warm temperate zone. The geographical distribution of fauna in Hubei Shennongjia belongs to the Oriental Realm, the China-India Subrealm, the Central China Region, and the Western Mountain Plateau Subregion. Faunal composition and elements from the Palaearctic and Oriental Realms blend in the site (Jiang et al., 2016b). The largest portions of terrestrial vertebrate fauna within Hubei Shennongjia are the Oriental type (28.3%) and the South-China type (17.9%). These fauna mixed with fauna of the Palaearctic type (13.7%), of the Himalaya-Hengduan Mountains type (10.9%), of the Northeast type (7.8%), and of the Holarctic type (5.2%). Birds and mammals belonging to the Oriental Realm dominate the fauna of Hubei Shennongjia (Table 4.1). They also show a blending of the faunal composition and elements of the Palaearctic region, revealing the ancient evolutionary origin of the biome and the diversity of wild animal species (Jiang, 2015).

4.2 Animal Species

A total of 629 vertebrate species (including 87 mammal species, 399 bird species, 53 reptile species, 37 amphibian species, and 53 fish species) are found in Hubei Shennongjia (Zhou et al., 2018). In addition, Hubei Shennongjia has 4365 insect species (belonging to 26 orders, 297 families, and 2227 genera) (Table 4.2). Hubei Shennongjia covers less than 0.01% of China, but it is an unsurpassed and wonderful animal paradise with rich diversity of mammals, birds, reptiles, amphibians, and

Z. Xie, G. Shen, *The Outstanding Universal Value and Conservation of Hubei Shennongjia*, https://doi.org/10.1007/978-981-16-0684-7_4

Table 4.1 Faunal composition of terrestrial vertebrates of Hubei Shennongjia

Type	Mammals	Birds	Reptiles	Amphibians	Total	Percentage (%)
Holarctic type (C)	2	28			30	5.2
Palaearctic type (U)	10	69			79	13.7
Northeast type (M)		45			45	7.8
North-China type (B)	2		1		3	0.5
Northeast-NC type (X)	1	5	1	3	10	1.7
Monsoon type (E)	7	4	8	4	23	4.0
Central-Asia type (D)		3			3	0.5
Plateau type (P)	1	4			5	0.9
Himalaya-Hengduan mountains type (H)	9	48	2	4	63	10.9
Yunnan–Guizhou plateau type (Y)			1	1	2	0.3
South-China type (S)	22	35	28	18	103	17.9
Oriental type (W)	27	118	12	6	163	28.3
Northern widely-distributed type (O)	6	40		1	47	8.2
Total species	87	399	53	37	576	100.0

Table 4.2 Wild vertebrate animals in Hubei Shennongjia

Class	Orders	Families	Genera	Species	Percentage (%)
Mammals	7	24	64	87	1.74
Birds	18	67	189	399	7.99
Reptiles	2	10	35	53	1.06
Amphibians	2	9	24	37	0.74
Fishes	4	12	42	53	1.06
Insects	26	297	2227	4365	87.41
Total	59	419	2581	4994	

insects in the north subtropics. The mammal species account for 14.13% and 71.9% of the state and Hubei Province, birds 31.39% and 76.58%, reptiles 13.15% and 67.95%, and amphibians 9.02% and 53.62%, respectively (Zhou & Lei, 2019). Comparing with adjacent Three Gorges areas, Chongqing Municipality and Jiangxi Province, Hubei Shennongjia contains 59.18–87.88% mammal species of these areas, birds 76.58%–96.14%, and amphibians 53.62%–86.05%. Number of reptiles in Hubei Shennongjia is more than adjacent Three Gorges areas, and also accounts for 84.13% and 100% of Chongqing Municipality and Jiangxi Province. All these evidences suggest that Hubei Shennongjia is a distribution center of animals beyond the adjacent region (Table 4.2).

4.2.1 Mammals

Based on surveys and existing scientific research, there are 87 mammal species in Hubei Shennongjia, including numerous rare and endangered species such as Sichuan snub-nosed monkey (*Rhinopithecus roxellana*), clouded leopard (*Neofelis nebulosa*), clouded leopard (*Panthera pardus*), Asian golden cat (*Catopuma temminckii*), dhole (*Cuon alpines*), Asian black bear (*Ursus thibetanus*), large Indian civet (*Viverra zibetha*), forest musk deer (*Moschus berezovskii*), Chinese goral (*Naemorhedus griseus*), and Chinese serow (*Capricornis milneedwardsii*), etc., (Jiang et al., 2016b; Zhou & Lei, 2019).

4.2.1.1 Sichuan Snub-Nosed Monkey (*Rhinopithecus roxellana*)

The Sichuan snub-nosed monkey (*Rhinopithecus roxellana*) is only distributed in Hubei, Sichuan, Shaanxi and Gansu provinces. The distribution areas of the three subspecies are isolated from each other, resulting in island-like distributions. Hubei Shennongjia is on the eastern edge of the distribution area of the Sichuan snub-nosed monkey. It is the distribution area of its Hubei subspecies. Within Hubei Shennongjia, snub-nosed monkeys are found in evergreen broadleaf forests, deciduous broadleaf forests, and conifer forests at elevations ranging from 1600 to 3000 m above sea level. According to the latest survey, there are 1550 golden monkeys in eight groups. They mainly live in Golden Monkey Ridge, Da Longtan, Xiao Longtan, Da Qianjiaping, and Xiao Qianjiaping, with an altitude ranging from 1600 to 3000 m. Statistics show that the main distribution area in Hubei Shennongjia is Golden Monkey Ridge sub-area (about 800 individuals), and Qianjianping sub-area (about 400 individuals) (Quan & Xie, 2002), there are a few small groups or individuals in Baiyanpo and Banbiyan (Photo 4.1). The Sichuan snub-nosed monkey is listed on the IUCN Red List (2014) as an endangered (EN) species. It is also included in the CITES Appendix I. Sichuan snub-nosed monkeys are key protected wild animals of Class I according to the *National Key Protected Wild Animals* which is the appendix to the *List of Wildlife Protection Act of People's Republic of China*. To achieve effective protection for the snub-nosed monkeys, the *Conservation and Research Center of the Sichuan Snub-nosed Monkey of Hubei Province* was established in December 2007 within Hubei Shennongjia. The State Forestry Administration of the People's Republic of China also approved the establishment of the first State Golden Monkey Research Center.

4.2.1.2 Asian Black Bear (*Ursus thibetanus*)

The Asian black bear (*Ursus thibetanus*) is a type of large animal belonging to the order Carnivora. Hubei Shennongjia is an original habitat of the Asian black bear and has the most concentrated distribution of a large population. Within Hubei

Photo 4.1 Sichuan snub-nosed monkey (*Rhinopithecus roxellana*)

Photo 4.2 Asian black bear (*Ursus thibetanus*)

Shennongjia, Asian black bears are found in areas from broadleaf and mixed forests at low altitude to coniferous forests up to 3000 m above sea level. They live at higher elevations in the summer, and descend during the winter. The Asian black bear is listed on the IUCN Red List (2014) as a vulnerable (VU) species. It is also included in the CITES Appendix I. Asian black bears are key protected wild animals of Class II according to the *National Key Protected Wild Animals* which is the appendix to the *List of Wildlife Protection Act of People's Republic of China* (Photo 4.2).

4.2.1.3 Forest Musk Deer (*Moschus berezovskii*)

The forest musk deer (*Moschus berezovskii*) is a small ungulate. Within Hubei Shennongjia, it is mainly distributed in broad-leafed forest, coniferous, and mixed forest at altitudes of between 1000 and 3000 m above sea level. It is listed on the IUCN Red List (2014) as an endangered (EN) species. It is also included in the CITES Appendix II. Forest musk deer are a key protected wild animal of Class I according to the *National Key Protected Wild Animals* which is the appendix to the *List of Wildlife Protection Act of People's Republic of China* (Photo 4.3).

4.2.1.4 Chinese Serow (Capricornis milneedwardsii)

The Chinese serow (*Capricornis milneedwardsii*) is a typical species of tropical and subtropical forests in Southeast Asia. In Hubei Shennongjia, the Chinese serow lives mainly in broadleaf forest, coniferous, and mixed forest at an altitude between 1000 and 3000 m. The Chinese serow is listed on the IUCN Red List (2014) as a Near Threatened (NT) species. It is also included in CITES Appendix I. Chinese serow is also designated by the *China Red Data Book of Endangered Animals* (Red Book) as a vulnerable (VU) species, and they are key protected wild animals of Class II according to the *National Key Protected Wild Animals* which is the appendix to the *List of Wildlife Protection Act of People's Republic of China* (Photo 4.4).

Photo 4.3 Forest musk deer (*Moschus berezovskii*)

Photo 4.4 Chinese serow (*Capricornis milneedwardsii*)

4.2.1.5 Chinese Goral (Naemorhedus griseus)

The Chinese goral (*Naemorhedus griseus*) is a typical species of subtropical and temperate forests in Asia. In Hubei Shennongjia, the Chinese goral lives mainly in evergreen forest, coniferous, and broadleaf forest. The Chinese goral is listed on the IUCN Red List (2014) as a vulnerable (VU) species. It is also included in the CITES Appendix I. Chinese goral are also designated by the *China Red Data Book of Endangered Animals* (Red Book) as an endangered (EN) species, and are key protected wild animals of Class II according to the *National Key Protected Wild Animals* which is the appendix to the *List of Wildlife Protection Act of People's Republic of China* (Photo 4.5).

4.2.2 Birds

Hubei Shennongjia is located in the transition region from the Oriental Realm to the Palaearctic Realm. It has complete altitudinal natural vegetation zones which provide various habitat types for birds from the Oriental and Palaearctic Realm. Based on recent surveys and existing scientific research, there are 399 bird species in Hubei Shennongjia. Hubei Shennongjia is also important migration routes for migratory birds, such as the great cormorant (*Phalacrocorax carbo*) and the Black-headed gull (*Larus ridibundus*). Hubei Shennongjia can provide safe migration channels with adequate food for these migratory birds (Yu et al., 2019; Zhang et al., 2016).

Photo 4.5 Chinese goral (*Naemorhedus griseus*)

4.2.2.1 Golden Eagle (Aquila chrysaetos)

The golden eagle (*Aquila chrysaetos*) is a large bird, belonging to the group of Falconiformes. It can be observed frequently in mountain forests and alpine grasslands, but is rarely seen over plains. Hubei Shennongjia provides the main forest habitats for golden eagles. The golden eagle is listed on the IUCN Red List (2014) as a world-class risk-free species, being considered as a less common (LC) species. It is included in the CITES Appendix II. Golden eagles are also designated by the *China Red Data Book of Endangered Animals* (Red Book) as a vulnerable (VU) species, and are key protected wild animals of Class I according to the *National Key Protected Wild Animals* which is the appendix to the *List of Wildlife Protection Act of People's Republic of China* (Photo 4.6).

4.2.2.2 Reeves's Pheasant (*Syrmaticus reevesii*)

The Reeves's pheasant (*Syrmaticus reevesii*) is an endemic species of pheasants in China. Within Hubei Shennongjia, it lives mainly in the evergreen and broadleaf forest. This species is listed on the IUCN Red List (2014) as a vulnerable (VU) species. Reeves's pheasants are also designated by the *China Red Data Book of Endangered Animals* (Red Book) as an endangered (EN) species. They are key protected wild animals of Class II according to the *National Key Protected Wild Animals* which is the appendix to the *List of Wildlife Protection Act of People's Republic of China* (Photo 4.7).

Photo 4.6 Golden eagle (*Aquila chrysaetos*)

Photo 4.7 Reevees's pheasant (*Syrmaticus reevesii*)

4.2.2.3 Koklass Pheasant (*Pucrasia macrolopha*)

The Koklass pheasant (*Pucrasia macrolopha*) is a species of the Pheasant family, the Phasianidae, and is the only species in the genus *Pucrasia*. Within Hubei Shennongjia, Koklass pheasants live mainly in coniferous and broadleaf forest. The Koklass pheasant is listed on the IUCN Red List (2014) as a world-class

Photo 4.8 Koklass pheasant (*Pucrasia macrolopha*)

risk-free species, being considered as a less common (LC) species. The birds are also designated by the *China Red Data Book of Endangered Animals* (Red Book) as a Near Threatened (NT) species. They are key protected wild animals of Class II according to the *National Key Protected Wild Animals* which is the appendix to the *List of Wildlife Protection Act of People's Republic of China* (Photo 4.8).

4.2.3 Reptiles

Based on recent surveys and existing scientific research, there are 53 species of reptiles in Hubei Shennongjia (Cai et al., 2016; Zhou & Lei, 2019).

4.2.3.1 Fea's Viper (*Azemiops feae*)

Fea's vipers (*Azemiops feae*) live mainly in the low hills and mountains at altitudes up to 1600 m in Hubei Shennongjia. Although it widely ranges over all of China's southern provinces and autonomous regions, the Fea's viper is a rare snake. It is the only species belonging to Azemiopinae, which represents the primitive group of Viperidae, thus having important scientific significance for tracing the origin and evolution of folding front snakes. The Fea's viper is listed on the IUCN Red List (2014) as a world-class risk-free species, being considered as a less common

Photo 4.9 Fea's viper (*Azemiops feae*)

(LC) species. Fea's vipers are also designated by the *China Red Data Book of Endangered Animals* (Red Book) as a vulnerable (VU) species (Photo 4.9).

4.2.3.2 Hundred-Pace Viper (Deinagkistrodon acutus)

The hundred-pace viper (*Deinagkistrodon acutus*) is an ancient species, which lives mainly in broadleaf forest at altitudes up to 1400 m in Hubei Shennongjia. It is listed on the IUCN Red List (2014) as a world-class risk-free species, being considered as a less common (LC) species. Hundred-pace vipers are also designated by the *China Red Data Book of Endangered Animals* (Red Book) as a vulnerable (VU) species. They are key protected wild animals of Class II according to the *National Key Protected Wild Animals* which is the appendix to the *List of Wildlife Protection Act of People's Republic of China* (Photo 4.10).

4.2.3.3 Striped Racer (*Elaphe taeniura*)

The striped racer (*Elaphe taeniura*) is a large snake species and its body length can be up to 2 m. It is a widely-distributed species, which lives mainly in broadleaf forest at altitudes up to 2400 m in Hubei Shennongjia. It is listed on the IUCN Red List (2014) as a world-class risk-free species, being considered as a less common (LC) species. Striped racers are also designated by the *China Red Data Book of Endangered Animals* (Red Book) as a vulnerable (VU) species (Photo 4.11).

Photo 4.10 Hundred-pace viper (*Deinagkistrodon acutus*)

Photo 4.11 Striped racer (*Elaphe taeniura*)

4.2.4 Amphibians

According to expedition records and existing scientific research, there are 37 species of amphibians in Shennongjia (Fei et al., 2012; Jiang et al., 2016b).

4.2.4.1 Chinese Giant Salamander (*Andrias davidianus*)

The Chinese giant salamander (*Andrias davidianus*) is the largest salamander and the largest amphibian in the world, reaching a length of 180 cm. It is endemic to the rocky, mountain streams and lakes in China. In Hubei Shennongjia, it is found in rocky hill streams at altitudes of 400–2000 m. The Chinese giant salamander is a relic species from the Quaternary glaciation. It is listed on both the IUCN Red List (2014) and in the *China Red Data Book of Endangered Animals* (Red Book) as a Critically Endangered (CR) species. It is also included in the CITES Appendix I. In 2008, the Chinese giant salamander was also listed as one of the top 10 "focal species" by the Evolutionarily Distinct and Globally Endangered (EDGE) project (Photo 4.12).

4.2.4.2 Wushan Salamander (*Ranodon shihi*)

The Wushan salamander (*Ranodon shihi*) shows interspecific disjunctive distribution with relative species from the Hynobiidae family. This reveals the impact of major geological and climatic events, such as the Himalayan orogeny, the uplift of the Tibetan Plateau and the Quaternary Glaciation, on their distribution pattern, and proves their ancient/relic features. Within Hubei Shennongjia, Wushan salamander lives mainly in the rocky streams of hills and mountains at elevations ranging from 1600 to 3000 m above sea level. The Wushan salamander is listed on both the IUCN Red List (2014) and in the *China Red Data Book of Endangered Animals* (Red Book) as a Near Threatened (NT) species (Photo 4.13).

Photo 4.12 Chinese giant salamander (*Andrias davidianus*)

Photo 4.13 Wushan salamander (*Ranodon shihi*)

Photo 4.14 Spiny-bellied frog (*Paa boulengeri*)

4.2.4.3 Spiny-Bellied Frog (*Paa boulengeri*)

The spiny-bellied frog (*Paa boulengeri*) is endemic to the rocky, mountain streams and lakes of China. In Hubei Shennongjia, it is found in rocky hill streams at altitudes of 700–1900 m. It is listed on the IUCN Red List (2014) as an endangered (EN) species. *Paa boulengeri* is also designated by the *China Red Data Book of Endangered Animals* (Red Book) as a vulnerable (VU) species (Photo 4.14).

4.2.5 Fishes

Based on recent surveys and existing scientific research, there are 53 species of fishes in Hubei Shennongjia. Of them, the holotype location of the Shennong goby (*Ctenogobius shennongensis* Yang & Xie, 1983) is in Hubei Shennongjia.

4.2.6 Insects

Based on recent surveys and existing scientific research, there are 4365 insect species (belonging to 26 orders, 297 families, and 2227 genera) in Hubei Shennongjia (Photo 4.15).

4.3 Rare and Endangered Animals

A total of 130 rare and endangered vertebrate species are found distributed in Hubei Shennongjia. Of them, 30 species are listed as endangered animals on the IUCN Red List (2016). In addition, 13 species are listed in Appendix I and 60 species are listed in Appendix II of the Convention on International Trade in Endangered Species of Wild Fauna and Flora (CITES) (2016) (Table 4.3) (Cai et al., 2016).

Photo 4.15 *Anisotima dispa*

Table 4.3 Rare and endangered animals in Hubei Shennongjia

Class	IUCN Red List (2016)	CITES (2016)		China species red list	National key protected wild animals list	
		I	II		I	II
Mammals	7	8	4	19	5	10
Birds	14	4	54	21	3	62
Reptiles	2	0	2	14	0	0
Amphibians	6	1	0	9	0	2
Fishes	1	0	0	10	0	1
Total	30	13	60	73	8	75

4.4 Endemic Animals

A total of 91 Chinese endemic vertebrate species are distributed in Hubei Shennongjia. Furthermore, Hubei Shennongjia is the distribution area of the Hubei subspecies of golden snub-nosed monkey. A total of 284 insect species are endemic to Hubei Shennongjia, accounting for 6.6% of the total insect species in Hubei Shennongjia.

4.5 Ancient/Relic Animal Species

The complex terrain in Hubei Shennongjia means it has been less affected by Quaternary Glaciation and has thus provided an ideal asylum for tertiary species throughout Eurasia. Hubei Shennongjia, therefore, has preserved a rich and intact ancient/relic species. These species include the Chinese giant salamander (*Andrias davidianus*), the Chinese hynobiid (*Hynobius chinensis*), the Wushan salamander (*Ranodon shihi*), and the Fea's viper (*Azemiops feae*). They also include varying monotypic families (families which only contain one genus one species), as well as mono- or oligo-genus.

4.6 Important Type Specimen Animal Hotspot

Hubei Shennongjia is the holotype location of 320 insect species, accounting for 7.4% of the total insect species in Hubei Shennongjia. Moreover, Hubei Shennongjia also contains the holotype location of the Shennong goby (*Ctenogobius shennongensis*) and the Hubei subspecies of golden snub-nosed monkey (*Rhinopithecus roxellana hubeiensis*) (Wu et al., 2019; Zhou et al., 2017b).

Chapter 5
Vegetation and Ecosystem

5.1 Biogeography

Under the biogeographic classification system established by Udvardy (1975), Hubei Shennongjia is located in the Palaearctic Realm of the world's eight biogeographical realms and belongs to the Oriental Deciduous Forest Biogeographical Province which is one of the 193 biogeographical provinces (Table 5.1 and Fig. 5.1).

5.2 Vegetation Types

According to the classification system of *Vegetation of China* (Editorial Committee of Vegetation Map of China, 2007), the natural vegetation in Hubei Shennongjia is classified into five vegetation groups, 11 vegetation types, and 46 formations (Table 5.2 and Photos 5.1, 5.2, 5.3, 5.4, 5.5, 5.6, 5.7, 5.8, 5.9, 5.10, 5.11, 5.12, 5.13, 5.14, 5.15, 5.16, 5.17, 5.18, 5.19, 5.20, and 5.21).

5.3 Vegetation Altitudinal Zonality

Hubei Shennongjia exhibits significant altitudinal zonation due to its diverse landforms, varied soil types and climate conditions. The diverse environments and habitats are suitable for various plants, so the species and vegetations are abundant. From the lower to the higher mountains, and from shady slopes to sunny slopes, the vegetation pattern and distribution show significant diversity pattern. The mountain vegetation belts begin with subtropical broadleaf evergreen forests at the baseband. Subtropical broadleaf evergreen forests cover all the low elevation valleys. The

Z. Xie, G. Shen, *The Outstanding Universal Value and Conservation of Hubei Shennongjia*, https://doi.org/10.1007/978-981-16-0684-7_5

Table 5.1 Location of Hubei Shennongjia according to Udvardy's biogeographic classification system

Biogeographical realm	Biogeographical province	Biome
Palaearctic realm	Oriental deciduous Forest	Temperate broadleaf forests

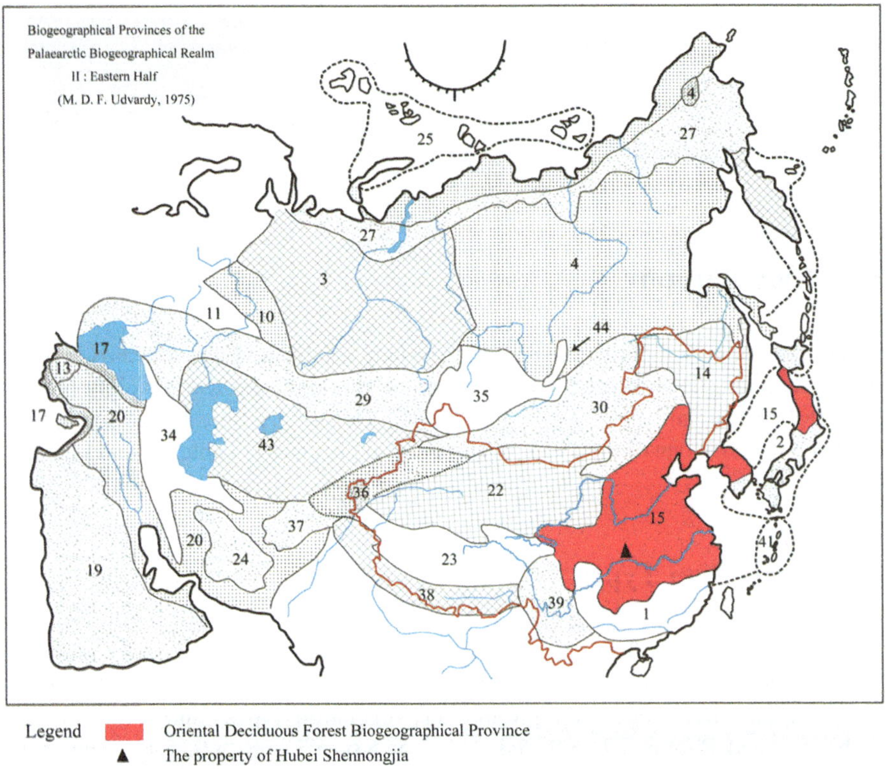

Fig. 5.1 Oriental deciduous forest biogeographical provinces (according to Udvardy, 1975)

altitudinal zonality of vegetation in Shennongjia is very distinct (Ma et al., 2017). In particular, the vegetation zonality on the Shennongjia's sunny slopes is more significant than that on its shady slopes, due to the fact that the sunny slopes are steeper. Regarding the distribution of vegetation in Hubei Shennongjia, mountain scrubs and subalpine scrubs cover 11.5%, meadows cover 5.0%, needleleaf forests cover 11.9%, needleleaf and broadleaf forests cover 29.6%, broadleaf forest cover 42%. Hubei Shennongjia has an area of 17,365 ha of Original forests, which cover 23.7% of the total area of Hubei Shennongjia.

The natural conditions of Hubei Shennongjia vary greatly with altitude. As a result, there are six typical zonalities in Hubei Shennongjia along the elevational gradients. The zonalities (Fig. 5.2) are subtropical evergreen broad-leaved forests,

Table 5.2 Vegetation types in Hubei Shennongjia Form

Vegetation groups	Vegetation types	Formations
I. Needleleaf Forest	1.Subtropical needleleaf forest	Form. *Pinus massoniana*
		Form. *Cunninghamia lanceolata*
		Form. *Pinus henryi*
		Form. *Pinus armandii*
	2.Subtropical and tropical mountains needleleaf forest	Form. *Abies chensiensis*
		Form. *Abies chensiensis, Picea wilsonii*
		Form. *Abies fargesii*
II. Mixed needleleaf and broadleaf Forest	1.Subtropical mountains mixed needleleaf, broadleaf evergreen and deciduous forest	Form. *Pinus armandii, Populus davidiana, Betula albosinensis*
		Form. *Pinus armandii, populus davidiana*
		Form. *Pinus armandii, Querus aliena* var. *acuteserrata*
		Form. *Pinup armandii, Betula utilis*
		Form. *Pinus massoniana, Quercus variabilis*
		Form. *Abies fargesii, Betula albosinensis, Acer* sp.
III. Broadleaf Forest	1. Subtropical broadleaf evergreen forest	Form. *Phoebe zhennan, Cyclobalanopsis gracilis*
		Form. *Cyclobalanopsis oxyodon, Sycopsis chinense, Quercus engleriana, Cyclobalanopsis myrsinifolia*
		Form. *Quercus spinosa*
	2.Subtropical mixed broadleaf evergreen and deciduous forest	Form. *Lindera communis, Cyclobalanopsis gracilis, Platycarya strobilacea, Betula luminifera*
		Form. *Quercus engleriana, Cyclobalanopsis oxyodon, Betula luminifera, Platycarya strobilacea*
		Form. *Quercus phillyraeoides, Q. acrodonta, Carpinus* sp., *Platycarya strobilacea*
	3.Subtropical broadleaf deciduous forest	Form. *Quercus glandulifera* var. *brevipetiolata*
		Form. *Quercus variabilis*
		Form. *Quercus variabilis, Q. aliena* var. *acuteserrata, Castanea seguinii*
		Form. *Quercus aliena* var. *acuteserrata*
		Form. *Fagus engleriana*
		Form. *Quercus aliena* var. *acuteserrata, Fagus engleriana, Betula albosinensis*

(continued)

Table 5.2 (continued)

Vegetation groups	Vegetation types	Formations
		Form. *Betula luminifera, Platycarya strobilacea, Carpinus* sp.
		Form. *Rhus verniciflua, Quercus acutidentata, Cornus controversa, Platycarya strobilacea*
		Form. *Juglans cathayensis*
		Form. *Castanea seguinii*
		Form. *Betula albosinensis*
IV. Scrub	1.Subtropical and tropical broad-leaf evergreen and deciduous scrub	Form. *Coriaria sinica, coggygria* var. *pubescens*
		Form. *Lespedeza formosa, L. buergeri*
		Form. *Quercus acrodonta*
		Form. *Salix caprea* var. *inica, Crataegus wilsonii, Sorbus hupehensis*
		Form. *Corylus heterophylla* var. *sutchuenensis, Viburnum opulus* var. *calvescens, Malus hupehensis*
		Form. *Chimonanthus praecox*
		Form. *Berberis dasystachya*
		Form. *Sinarundinaria nitida*
	2.Subalpine broadleaf evergreen sclerophyllous scrub	Form. *Rhododendron fargesii*
	3.Subalpine needleleaf evergreen scrub	Form. *Sabina pingii* var. *wilsonii*
	4.Subalpine broadleaf deciduous scrub	Form. *Cotoneaster horizontalis*
		Form. *Salix cupularis*
V. Meadow	1.Temperate grass and forb meadow	Form. *Miscanthus sinensis, Pteridium aquilinum* var. *latiusculum*
		Form. *Trisetum clarkei, Festuca rubra, Deyeuxia scabrescens*
		Form. *Carex* sp., *Sanguisorba officinalis* var. *longifolia, Anaphalis sinica, Geranium henryi*
		Form. *Carex* sp., *Juncus concinnus, Sanguisorba officinalis* var. *longifolia, Chamaenerion angustifolium*

northern subtropical mixed evergreen and deciduous broad-leaved forests, warm temperate deciduous broad-leaved forests, temperate mixed needle-leaved and broad-leaved forests, cold temperate coniferous forests, and subalpine fruticose and meadow communities (Ma et al., 2017).

Photo 5.1 Form. *Abies fargesii*

5.3.1 Subtropical Evergreen Broad-Leaved Forest Belt

The subtropical evergreen broad-leaved forest ecosystems in Hubei Shennongjia are the baseband of the altitudinal zonality with an elevation of less than 1000 m. The dominant trees of these forests are most from beech family, including blue Japanese oak (*Cyclobalanopsis glauca*), sharpdent oak (*Cyclobalanopsis oxyodon*), myrsinaleaf oak (*Cyclobalanopsis myrsinifolia*), engler oak (*Quercus engleriana*), oriental oak (*Quercus variabilis*) and faber oak (*Quercus fabrei*), and phoebe (*Phoebe* sp.), littlefuite machilus (*Machilus microcarpa*), Chinese fighazel (*Sycopsis sinensis*), etc. (Xie et al., 2020). The common animals in these forests are the rhesus monkey (*Macaca mulatta*), the Malayan porcupine (*Hystrix brachyura*), the common magpie (*Pica pica*), the red-billed blue magpie (*Urocissa erythroryncha*), the Eurasian tree sparrow (*Passer montanus*), the black drongo (*Dicrurus macrocercus*), the ashy drongo (*Dicrurus leucophaeus*), the barn swallow (*Hirundo rustica*), the red-rumped swallow (*Hirundo daurica*), the gray-headed woodpecker (*Picus canus*), spotless tree frog (*Hyla immaculata*), etc. The soil of the forest is yellow-cinnamon soil (Photo 5.22) (Ge et al., 2012; Xu et al., 2017)

Photo 5.2 Form. *Quercus engleriana, Cyclobalanopsis oxyodon, Betula luminifera, Platycarya strobilacea*

Photo 5.3 Form. *Pinus henryi*

Photo 5.4 Form. *Quercus spinosa*

Photo 5.5 Form. *Rhododendron fargesii*

Photo 5.6 Form. *Betula albosinensis*

Photo 5.7 Form. *Betula utilis*

Photo 5.8 Form. *Pinus armandii, Quercus aliena* var. *acuteserrata*

Photo 5.9 Form. *Pinus armandii, Populus davidiana, Betula albosinensis*

Photo 5.10 Form. *Sinarundinaria nitida*

Photo 5.11 Form. *Chimonanthus praecox*

Photo 5.12 Form. *Betula luminifera, Platycarya strobilacea, Carpinus* sp.

Photo 5.13 Form. *Coriaria nepalensis, Cotinus coggygria* var. *pubescens*

Photo 5.14 Form. *Cyclobalanopsis oxyodon, Sycopsis sinensis, Quercus engleriana, Cyclobalonopsis myrsinifolia*

Photo 5.15 Form. *Fagus engleriana*

Photo 5.16 Form. *Cotoneaster horizontalis*

Photo 5.17 Form. *Quercus aliena* var. *acuteserrata*

Photo 5.18 Form. *Quercus variabilis*

Photo 5.19 Form. *Quercus phillyraeoides, Quercus acrodonta, Carpinus* sp., *Platycarya strobilacea*

Photo 5.20 Form. *Toxicodendron succedaneum, Quercus aliena* var. *acutiserrata, Cornus controversa, Platycarya strobilacea*

Photo 5.21 Form. *Trisetum clarkei*, *Festuca rubra*, *Deyeuxia scabrescens*

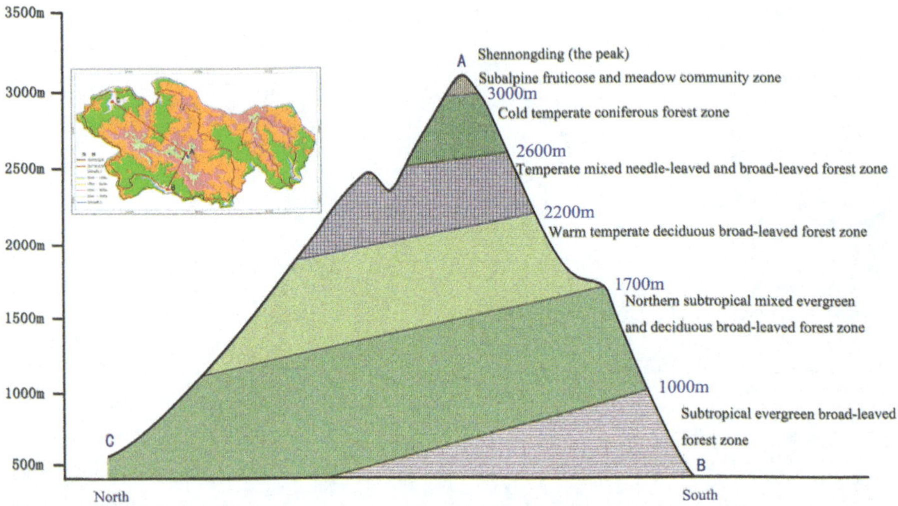

Fig. 5.2 Altitudinal zonality of the vegetation in Hubei Shennongjia

Photo 5.22 Subtropical evergreen broad-leaved forest belt

5.3.2 *Northern Subtropical Mixed Evergreen and Deciduous Broad-Leaved Forest Belt*

The mixed evergreen and deciduous broad-leaved forests are the typical forests of elevations ranging from 1000 to 1700 m. The dominant formation of these forests is engler beech (*Fagus engleriana*)—manyvei oak (*Cyclobalanopsis multinervis*), farges clethra (*Clethra fargessi*)—Henry tanoak (*Lithocarpus henryi*), dove tree (*Davidia involucrata*)—myrsinaleaf oak (*Cyclobalanopsis myrsinifolia*), true lacquertree (*Toxicodendron vernicifluum*)—myrsinaleaf oak (*Cyclobalanopsis myrsinifolia*), sharpdent oak (*Cyclobalanopsis oxyodon*)—dyetree (*Platycarya strobilacea*) (Xie et al., 2020). The common animals in these forests are the Asian black bear (*Ursus thibetanus*), the wild boar (*Sus scrofa*), the Chinese bamboo rat (*Rhizomys sinensis*), the Confucian niviventer (*Niviventer confucianus*), the Chinese sparrowhawk (*Accipiter soloensis*), the Jerdon's baza (*Aviceda jerdoni*), the peregrine falco (*Falco peregrinus*), the Eurasian hobby (*Falco subbuteo*), the Asian emerald cuckoo (*Chalcites maculatus*), west China toad (*Bufo gargarizans*), etc. The soil of these forests is yellow brown soil (Photo 5.23).

Photo 5.23 Northern subtropical mixed evergreen and deciduous broad-leaved forest belt

5.3.3 Warm Temperate Deciduous Broad-Leaved Forest Belt

The warm temperate deciduous broad-leaved forests are typical forests in the elevation range of 1700–2200 m. The dominant formations of these forests are short stipes oak(*Quercus serrata* var. *brevipetiolata*), seguin chestnut(*Castanea seguinii*), Henry chestnut (*Castanea henryi*), sharptooth oak (*Quercus aliena* var. *acutiserrata*), shiningleaf birch (*Betula luminifera*), Himalayan birch (*Betula utilis*), Chinesepaper birch (*Betula albosinensis*), engler beech, Taiwan beech, David poplar (*Populus davidiana*), retuse ash (*Fraxinus insularis*), dyetree (*Platycarya strobilacea*), Chinese walnut (*Juglans cathayensis*), manyseeded euptelea (*Euptelea pleiosperma*), giant dogwood (*Cornus controversa*)and true lacquertree (*Toxicodendron vernicifluum*), roundwing fruit cycloca(*Cyclocarya paliurus*), etc. (Fan et al., 2017a). The common animals in these forests are the Chinese goral (*Nemorhedus griseus*), the Père David's rock squirrel (*Sciurotamias davidianus*), the leopard cat (*Prionailurus bengalensis*), the golden pheasant (*Chrysolophus pictus*), Chinese brown frog (*Rana chensinensis*), etc. The soil of these forests is mountain brown soil.

5.3.4 Temperate Mixed Needle-Leaved and Broad-Leaved Forest Belt

Temperate mixed needle-leaved and broad-leaved forests are typical forests between elevation of 2200–2600 m. The dominant formations of these forests are farges fir (*Abies fargesii*)—past over thododendron (*Rhododendron praeteritum*), farges fir (*Abies fargesii*)—maple (*Acer* sp.), Chinesepaper birch (*Betula albosinensis*)—farges fir (*Abies fargesii*), Chinesepaper birch (*Betula albosinensis*)—armand pine (*Pinus armandii*), armand pine (*Pinus armandii*)—David poplar (*populus davidiana*), armand pine (*Pinus armandii*)—wallich willow (*Salix wallichiana*), hornbeam (*Carpinus* sp.)—Chinese hemlock (*Tsuga chinensis*), etc. The common animals of these forests are the Sichuan snub-nosed monkey (*Rhinopithecus roxellana*), the Chinese serow (*Capricornis milneedwardsii*), the Asian golden cat (*Catopuma temminckii*), the golden eagle (*Aquila chrysaetos*), the Chinese sparrowhawk (*Accipiter gularis*), Chinese luehdorfia (*Luehdorfia chinensis*), etc. The soil for these forests is mountain dark brown soil (Photo 5.24).

5.3.5 Cold Temperate Coniferous Forest Belt

The cold temperate coniferous forests are typical forests in the elevation of 2600–3000 m. The dominant formations of the forests are farges fir. The common animals of these forests are the Sichuan snub-nosed monkey (*Rhinopithecus*

Photo 5.24 Temperate mixed needle-leaved and broad-leaved forest belt

roxellana), the forest musk deer (*Moschus berezovskii*), the white-bellied green pigeon (*Treron sieboldii*), the Himalayan swiftlet (*Aerodramus brevirostris*), the gray-capped woodpecker (*Dendrocopos canicapillus*), the long-tailed minivet (*Pericrocotus ethologus*), the spotted nutcracker (*Nucifraga caryocatactes*), the common rosefinch (*Carpodacus erythrinus*), gray-headed bullfinch (*Pyrrhula erythaca*), etc. The soil of these forests is brown coniferous forest soil (Photo 5.25).

5.3.6 Subalpine Fruticose and Meadow Community Belt

Subalpine fruticose and meadow is the typical community at an elevation of 3000 m. The dominant formations of these communities are Pere farges rhododendron (*Rhododendron fargesii*), Wilson juniper (*Juniperus pigii* var. *wilsonii*), fountain bamboo (*Sinarundinaria nitida*), clark trisetum (*Trisetum clarkei*), red fescuegrass (*Festuca rubra*), and scabrous small reed (*Deyeuxia scabrescens*). The common animals of these communities are the forest musk deer (*Moschus berezovskii*), the Chinese serow (*Capricornis milneedwardsii*), the Chinese bamboo partridge (*Bambusicola thoracica*), Eurasian hoopoe (*Upupa epops*), etc. The soil of these communities is meadow brown soil (Photo 5.26).

Photo 5.25 Cold temperate coniferous forest belt

Photo 5.26 Subalpine fruticose and meadow community belt

5.4 Habitats

Hubei Shennongjia's unique geographical and climatic features incubate a richness of biodiversity and ecosystems. According to the IUCN/SSC categorization of global habitats, there are six kinds of first-level IUCN/SSC habitats in Hubei Shennongjia. The main habitats such as forests, shrublands and grasslands, which form diverse terrestrial ecosystems were discussed before (Table 5.3).

Wetland ecosystem, rock barren areas, and caves are important habitats in Hubei Shennongjia. Hubei Shennongjia is the headstream of many streams of the Yangtse and Han Rivers, which support lots of river wetlands that are the habitats for many fish and amphibians, such as *Andrias davidianus* (Photo 5.27).

There are large areas of rocklands with limestone texture. The rocklands are dominated by moss, lichen and small vascular plants (Photo 5.28).

There are lots of limestone caves in Hubei Shennongjia. The common species in the caves are the Edward's leopoldamys (*Leopoldamys edwardsi*), the red and white giant flying squirrel (*Petaurista alborufus*), the brown dipper (*Cinclus pallasii*), horseshoe bats (*Rhinolophus* spp.), the Hodgson's myotis (*Myotis formosus*), the Himalayan swiftlet (*Aerodramus brevirostris*), red-spotted toothed toad (*Oreolalax rhodostigmatus*), etc. (Photo 5.29).

Table 5.3 First-level IUCN/SSC habitats in Hubei Shennongjia (IUCN/SSC, 2011)

First-level IUCN/SSC habitats	Property
1. Forest	√
2. Savannah	
3. Shrubland	√
4. Grassland	√
5. Wetlands (inland)	√
6. Rocky barren areas	√
7. Caves and subterranean	√
8. Desert	
9. Sea	
10. Coastline/intertidal	
11. Artificial–terrestrial	
12. Artificial–aquatic	
13. Introduced vegetation	

Photo 5.27 Wetland ecosystem

Photo 5.28 Rockland ecosystem

Photo 5.29 Cave ecosystem

Chapter 6
Outstanding Universal Value

Hubei Shennongjia has an area of 73,318 ha, and its buffer zones have an area of 41,536 ha. The heritage is situated on the transitional zone between China's second level and the third levels of terrain, which is also the key water conservation area for Yangtze River and Han River. Dominated by Subtropical Gyre, the climate of Shennongjia is the transition zone from subtropical to warm temperate. This unique geographical environment makes Hubei Shennongjia an outstanding example among the mountains of similar latitude across the world, in terms of its richful plants and animals and their dynamic process. Hubei Shennongjia provides a case for typical elevation vegetation in the Oriental Deciduous Forest biogeographical province. Hubei Shennongjia is a vital source of global temperate flora and it has the highest concentration of global temperate genera (Chen et al., 2019). It is also an outstanding representative of biological evolution in the Oriental Deciduous Forest biogeographical province.

Hubei Shennongjia has exceptional biodiversity and is a key refuge for biodiversity. The number of deciduous woody species in Hubei Shennongjia is the richest in the world (Fan et al., 2017a). These species are of outstanding and universal protective and scientific value. Hubei Shennongjia is a prominent case of ongoing ecosystem dynamics in the succession of subtropical mixed broad-leaved evergreen and deciduous forest in the Northern Hemisphere. Hubei Shennongjia offers an example of a special case for elevational vegetation in the Oriental Deciduous Forest biogeographical province. It is also an outstanding example for research into the ecological processes of mountain ecosystem at different elevation gradients under global climate change.

Z. Xie, G. Shen, *The Outstanding Universal Value and Conservation of Hubei Shennongjia*, https://doi.org/10.1007/978-981-16-0684-7_6

6.1 Ecological and Biological Value

Hubei Shennongjia is a prominent case representing ongoing ecosystem dynamics in the succession of natural subtropical mixed broad-leaved evergreen and deciduous forest in the Northern Hemisphere. It provides an example of typical mountain altitudinal biological zones in the Oriental Deciduous Forest biogeographical province. It is also an outstanding example for research on the ecological processes of mountain ecosystems along the elevation gradients under global climate change. Hubei Shennongjia is also a vital source of global temperate flora and it has the highest concentration of global temperate genera (Xie et al., 2017). It is an outstanding representative of biological evolution in the Oriental Deciduous Forest biogeographical province (Xie & Shen, 2018). The unique terrain and climate conditions, together with the fact that it was less affected by the Quaternary Glaciation, have all made Hubei Shennongjia an ideal habitat for numerous relic, rare, endangered, endemic, type, and deciduous woody species, which are of outstanding and universal protective and scientific value.

6.1.1 Outstanding Examples of Subtropical Mixed Evergreen and Deciduous Broadleaf Forest

In the Chinese subtropics, because of the effects of relative sea-land positions, as well as the general circulation and existence of the Qin–Tibet plateau, the monsoon subtropics affect such a great proportion of the land, meaning that there is an extensive area with a moist warm subtropical climate. However, there is not desert vegetation or Mediterranean broadleaf evergreen sclerophyllous forests in the same latitude zone controlled by subtropical high pressure. Instead, subtropical broadleaf evergreen forests have widely covered the Chinese subtropics, becoming an unusual and rare type of vegetation on the earth (Fig. 6.1) (Zhou et al., 2019). Mixed broadleaf evergreen and deciduous forests, which are the transitional vegetation types from warm temperate broadleaf deciduous forests to mid-subtropical broadleaf evergreen forests, are the typical vegetation type of the north subtropical zonal vegetation (Table 6.1). Mixed broadleaf evergreen and deciduous forests are forests consisting of evergreen broad-leaved species of *Cyclobalanopsis, Castanopsis*, and *Lithocarpus*, and deciduous broad-leaved species of *Quercus* and *Fagus*.

Hubei Shennongjia is located on Qinba Mountain, belonging to the "Daba Mountain evergreen forests" subcategory, which falls under the category of "Southwest China Temperate Forests" in the Global 200 categorization of world ecoregions formed by WWF. Hubei Shennongjia is a typical representative of Daba Mountain evergreen forests due to its subtropical mixed ecological system and well protected condition (Fig. 6.2), where the typical vegetation is mixed evergreen and deciduous broadleaf forests. Hubei Shennongjia has preserved the most intact mixed broadleaf evergreen and deciduous forests in China and even in the Northern Hemisphere.

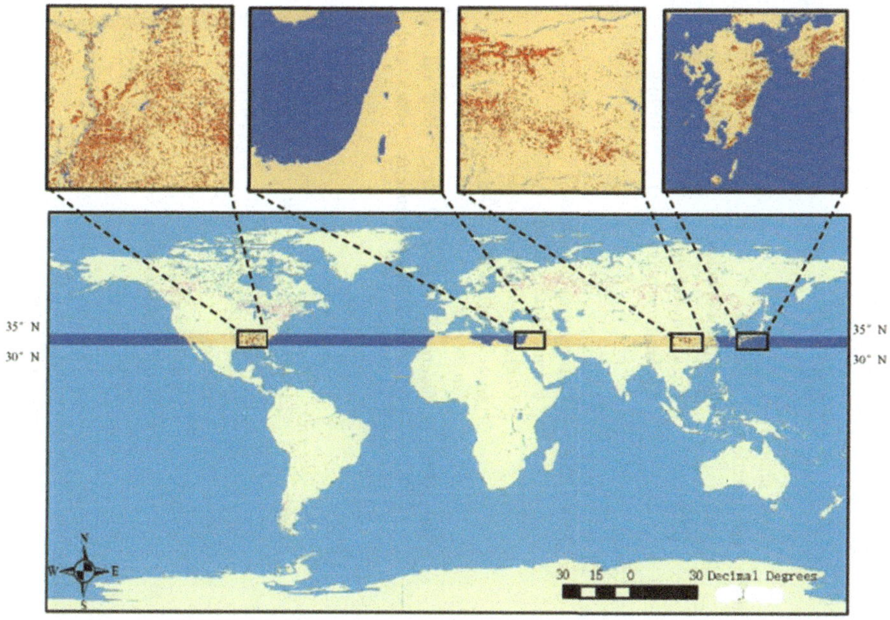

Fig. 6.1 Evergreen deciduous broad-leaved mixed forests (red areas) along the belt of 30°–35°N in the Northern Hemisphere

Hubei Shennongjia, therefore, offers a typical demonstration model of the mixed broadleaf evergreen and deciduous forests (Xie et al., 2017; Zhou et al., 2019) in northern hemisphere. These forests are the fundamental links between warm temperate broadleaf deciduous forests and subtropical broadleaf evergreen forests. The presence of these forests here means that China has the most complete continuum of zonal forests, with forests ranging from cold temperate needleleaf forests to tropical rain forests.

6.1.2 The Typical Mountain Altitudinal Biological Zones

Within Hubei Shennongjia, the topography with a vertical elevation of between 400 and 3100 m was saved from the damaging effects of the Quaternary glaciers and the blast of cold wave from Mongolia-Siberia. This area has also benefitted from the southeast monsoon and southwest monsoon. Hubei Shennongjia is rich in flora due to the fact that it is a convergence zone between subtropical flora and warm temperate flora (Ge et al., 2013). The elevation natural vegatation includes subtropical evergreen broad-leaved forests, northern subtropical mixed evergreen and deciduous broad-leaved forests, warm temperate deciduous broad-leaved forests, temperate mixed needle-leaved and broad-leaved forests, cold temperate coniferous

Table 6.1 Overview of Shennongjia and its counterparts in the same latitude belt (Ma et al., 2017)

Country	Area	Vegetation type	Location	Altitude (m)	Mean annual temperature (°C)	Mean annual precipitation (mm)	Dominant species	Soil type
China	Shennongjia	Evergreen deciduous broad-leaved mixed forest	31.27°N, 110.48°E	1670	10.6	1200	*Fagus engleriana, Cyclobalanopsis glauca*	Mountain yellow brown soil
Japan	The southern lowlands of Honshu, north part of Shikoku, Kyushu, and Ryuryu	Evergreen broad-leaved forest	26.58°–35.5°N, 130°–140°E	250–1100	13–21	1300–2500	*Castanopsis cuspidata, C. cuspidata* var. *sieboldii, Quercus salicina, Q. acuta, Q. sessilifolia, Machilus thunbergii*	Brown soil
France, Italy, Spain, and Greece	Mediterranean	Sclerophyllous evergreen broad-leaved forest	30.25°–45.83°N, 0°–30°E	300–2800	16–20	300–1000	*Quercus coccifera, Q. ilex, Q. suber, Pistacia lentiscus*	Red soil and yellow soil
USA	Great Smoky Mountains	Temperate broad-leaved forest	35.6°N, 83.48°W	1830	12	1270–2030	*Quercus borealis, Tilia heterophylla, Betula alleghaniensis, Acer rubrum*	Mountain brown soil
North Africa	North of the Sahara Desert	Desert	25°–35°N, 15°W–50°E	–	>30	90–450	*Acacia, Artemisia, Aristida, Eragrostis, Panicum*	Rock desert (rock desert), gravel desert, and desert

Information in the table is from UNESCO official website (http://whc.unesco.org) and related published literature, e.g., He (2006).

Fig. 6.2 Southwest China temperate forests ecoregion

forests, and subalpine fruticose and meadow communities (Ma et al., 2017). Hubei Shennongjia is the key area in the shifting, exchanging, mixing, and evolution of the flora from Sino-Himalaya and Sino-Japanese. It is the high concentration of all of the subtropical, warm temperate, temperate, and cold temperate ecosystems in such a small area that makes Hubei Shennongjia an outstanding example for research on the ecological processes of mountain ecosystems along the elevation gradients under global climate change (Xie & Shen, 2018).

The vertical vegetation zonation in Shennongjia Natural Heritage shows an ecological process of gradual succession of dominant tree species and amounts with the elevation increasing (Ma et al., 2017). The southern slope of the Shennongjia Natural Heritage is the northern edge of the middle subtropical zone. The dominant species in the zone of evergreen broad-leaved forest are Fagaceae like *Cyclobalanopsis glauca, C. oxydon, C. myrsinifolia*, and *Quercus engleriana*. With the increase of altitude, some of the dominant species are gradually replaced by deciduous tree species like *Q. aliena* var. *acutiserrata and Fagus engleriana*, forming a zone of broad-leaved mixed evergreen and deciduous forest. With the further increase of altitude, the dominant evergreen tree species gradually decrease and deciduous tree species like *Quercus acuteserrata* and *Cyclobalanopsis glauca* become the dominant species, forming a zone of broad-leaved deciduous forest with deciduous trees as the dominant species (Ma et al., 2017). When the altitude rises to 2200–2600 m, coniferous species appear in the zone of broad-leaved deciduous forest with deciduous species as the dominant species. Zone of mixed broadleaf–conifer forest gradually takes the place of zone of broad-leaved deciduous forest as the altitude rises, with *Abies fargesii* and *Betula albosinensis* as dominant species. When the altitude rises above 2600 m, coniferous species take the place of *Betula albosinensis* in zone of broadleaf–conifer forest with *Abies fargesii* and *Betula albosinensis* as dominant species, forming a zone of cold temperate coniferous forest with *Abies fargesii* as the dominant species. When the altitude rises above 3000 m, the dominant tree species Abies fargesii are gradually replaced by *Rhododendron hypoglaucum*, *Sabina Squamata* var. *fargesii*, *Fargesia spathacea*, and *Arundinella hirta*, forming zone of mountain shrub meadow (Fig. 6.3).

By comparing and analyzing the vertical vegetation zonation in Shennongjia Natural Heritage with world natural heritage and tentative heritage in Oriental Deciduous Forest Biogeographic Province we have found that the elevational range of Shennongjia Natural Heritage is 2700 m, forming a complete vertical vegetation zonation in mountain. From bottom to top, there are zone of broad-leaved evergreen forest (at sun slope of the Heritage), zone of broad-leaved forest mixed evergreen and deciduous, zone of broad-leaved deciduous forest, zone of mixed broadleaf–conifer forest, zone of coniferous forest and zone of mountain shrub meadow.

Compared with the elevational biological zonation along the elevational gradient in the heritage, there is no zone of temperate broadleaf–conifer forest or zone of cold temperate coniferous forest in Huangshan (in Anhui Province) or Sanqingshan (in Jiangxi Province). Huangshan pine forests distributed at 800–1800 m in Huangshan and 1400–1700 m in Sanqingshan are warm to moderate temperate coniferous forest. There is no zone of temperate mixed broadleaf–conifer forest or zone of cold temperate coniferous forest in Badagongshan (in Hunan Province). The vertical vegetation zonation in Fanjingshan is a mixed broadleaf–conifer forest with *Abies fargesii* as the dominant species, which is distributed in a narrow altitude range of 2100–2350 m. There is no zone of cold temperate coniferous forest in the vertical vegetation zonation in Fanjingshan. The zonal vegetation type of Shirakami-Sanchi in Japan is broad-leaved deciduous forest, there is no zone of broad-leaved evergreen forest and broad-leaved mixed evergreen and deciduous forest in the vertical vegetation zonation in Shirakami-Sanchi. The zonal vegetation type of Mt. Fuji in Japan is only temperate broad-leaved deciduous forest rather than the evergreen broad-leaved forest, the mixed evergreen and deciduous broad-leaved forest, and mixed broadleaf–conifer forest (Ohsawa, 1984). The zonal vegetation type of Jiuzhaigou in Sichuan Province is coniferous broad-leaved mixed forest. There is no zone of broad-leaved deciduous forest, zone of mixed evergreen and deciduous broad-leaved forest and zone of evergreen broad-leaved forest in the vertical vegetation zonation in Jiuzhaigou. Compared with Shennongjia Natural Heritage, the vertical vegetation zonation of Taibaishan in Qinling Mountains is also relatively complete. However,

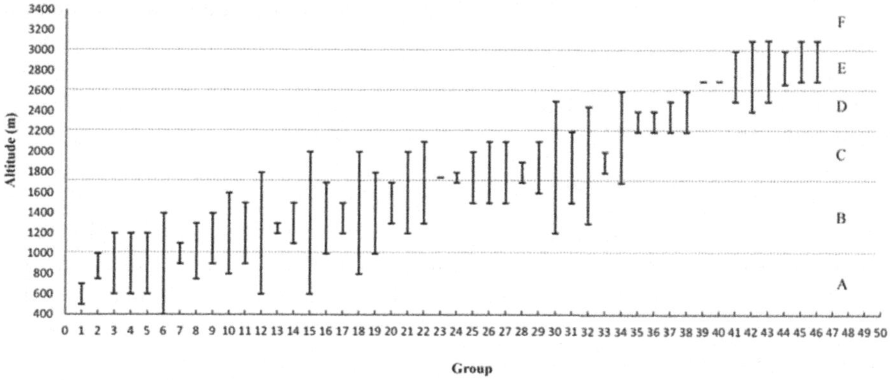

Fig. 6.3 The distribution of main plant formations in Shennongjia along the elevation gradient. Capital English letters represent vertical vegetation zones: **(a)** subtropical zone of evergreen broad-leaved forest; **(b)** north subtropical zone of Mixed evergreen and deciduous broad-leaved forest; **(c)**, warm temperate zone of broad-leaved deciduous forest; **(d)** temperate zone of mixed broadleaf–conifer forest; **(e)** cold temperate zone of coniferous forest; **(f)**, subalpine zone of shrub meadow. The abscissa represents different formations: 1, Form. *Chimonanthus praecox;* 2, Form. *Phoebe zhennan, Cyclobalanopsis gracilis;* 3, Form. *Coriaria sinica, Cotinus coggygria* var. *pubescens;* 4, Form. *Pinus massoniana, Quercus variabilis;* 5, Form. *Cunninghamia lanceolata;* 6, Form. *Quercus variabilis;* 7, Form. *Lindera communis, Cyclobalanopsis gracilis, Platycarya strobilacea, Betula luminifera;* 8, Form. *Pinus massoniana;* 9, Form. *Quercus acrodonta;* 10, Form. *Cyclobalanopsis oxyodon, Sycopsis sinensis, Quercus engleriana, Cyclobalanopsis myrsinifolia;* 11, Form. *Quercus phillyraeoides, Q. acrodonta, Carpinus* sp., *Platycarya strobilacea;* 12, Form. *Juglans cathayensis;* 13, Form. *Quercus variabilis, Q. aliena* var. *acuteserrata, Castanea seguinii;* 14, Form. *Quercus engleriana, Cyclobalanopsis oxyodon, Betula luminifera, Platycarya strobilacea;* 15, Form. *Quercus spinosa;* 16, Form. *Quercus glandulifera* var. *brevipetiolata;* 17, Form. *Castanea seguinii;* 18, Form. *Pinus henryi;* 19, Form. *Betula luminifera, Platycarya strobilacea, Carpinus* sp.; 20, Form. *Pinus armandii, Betula utilis;* 21, Form. *Quercus aliena* var. *acuteserrata;* 22, Form. *Abies chensiensis;* 23, Form. *Corylus heterophylla* var. *sutchuenensis, Viburnum opulus* var. *calvescens, Malus hupehensis;* 24, Form. *Carex* sp., *Sanguisorba officinalis* var. *longifolia, Anaphalis sinica, Geranium henryi;* 25, Form. *Rhus verniciflua, Quercus acutidentata, Cornus controversa, Platycarya strobilacea;* 26, Form. *Miscanthus sinensis, Pteridium aquilinum* var. *latiusculum;* 27, Form. *Lespedeza formosa, L. buergeri;* 28, Form. *Carex* sp., *Juncus concinnus, Sanguisorba officinalis* var. *longifolia, Chamaenerion angustifolium;* 29, Form. *Pinus armandii, Quercus aliena* var. *acuteserrata;* 30, Form. *Pinus armandii;* 31, Form. *Fagus engleriana;* 32, Form. *Abies chensiensis, Picea wilsonii;* 33, Form. *Quercus aliena* var. *acuteserrata, Fagus engleriana, Betula albosinensis;* 34, Form. *Betula albosinensis;* 35, Form. *Pinus armandii, Populus davidiana, Betula albosinensis;* 36, Form. *Pinus armandii, Populus davidiana;* 37, Form. *Salix caprea* var. *inica, Crataegus wilsonii, Sorbus hupehensis;* 38, Form. *Abies fargesii, Betula albosinensis, Acer* sp.; 39, Form. *Salix cupularis;* 40, Form. *Berberis dasystachya;* 41, Form. *Sinarundinaria nitida;* 42, Form. *Cotoneaster horizontalis;* 43, Form. *Abies fargesii;* 44, Form. *Rhododendron fargesii;* 45, Form. *Sabina pingii* var. *wilsonii;* 46, Form. *Trisetum clarkei, Festuca rubra, Deyeuxia scabrescens* (Ma et al., 2017)

its zone of mixed evergreen and broad-leaved deciduous forest has been seriously damaged due to long-term human activities. At present, the remained mixed evergreen and deciduous broad-leaved forest in Qinling is not intact. The remained vegetation is mainly *Pinus massoniana* forest and *Quercus acutissima* forest. The results show that Shennongjia Natural Heritage concentrates the typical vegetation types of subtropical, warm temperate, middle temperate, and cold temperate zone in a small horizontal distance. Shennongjia Natural Heritage also preserves a complete vertical vegetation zonation. Shennongjia Natural Heritage is unique and representative in the Oriental Deciduous Forest Biogeographic Province (Ma et al., 2017).

With the increase of altitude, the temperature, air pressure, and surface area decrease in mountains, and environmental conditions change significantly (Korner, 2007). The vertical vegetation zonation is the result of combined action of biotic and abiotic factors along the altitudinal gradient in mountains. The altitudinal gradient, with a smaller geographical scale, presents a larger climate gradient under similar environmental conditions and concentrates different vegetation types. It is a concentrated exhibition for many bioecological processes and a typical case for studying the changes of bioecological processes with climate (Malhi et al., 2010). The Qinghai–Tibet Plateau has caused the formation of the large-scale subtropical evergreen broad-leaved forest in the East Asia subtropical zone. It is not only different from the drought and heat tolerant sclerophyllous evergreen forest of Mediterranean type, but also different from the subtropical and tropical desert vegetation in the northern hemisphere at the same latitude (Editorial Committee of Vegetation Map of China, 2007). Shennongjia Natural Heritage extends across the middle and north subtropics and breeds the typical zonal mixed evergreen and deciduous broad-leaved forest in the northern hemisphere. Also, the altitudinal gradient in Shennongjia Natural Heritage concentrates the climate types of middle subtropics, north subtropics, and warm temperate, temperate, and cold temperate zones within a small horizontal distance and breeds a complete vertical vegetation zonation of the Oriental Deciduous Forest Biogeographic Province in the Classification of the Biogeographical Provinces of the World by Udvardy. It has formed evergreen broad-leaved forest, mixed evergreen and deciduous broad-leaved forest, broad-leaved deciduous forest, mixed broadleaf–conifer forest, coniferous forest, and mountain shrub meadow from low altitude to high altitude, showing the typical bioecological process of Oriental Deciduous Forest Biogeographic Province.

The intact elevation vegetation belts in the elevation gradients in the heritage shows the succession process of dominant species and functional groups of plant communities with the change of altitudinal gradient in the Oriental Deciduous Forest Biogeographic Province by Udvardy. Studies have found that the altitudinal vegetation gradients in mountain show the succession process of dominant species and functional groups of plant communities with the change of altitudinal gradient (Sundqvist et al., 2013), which is an ideal place to study the change of functional traits with climate. The dominant species of evergreen broad-leaved forest in Shennongjia Natural Heritage are evergreen tree species of Fagaceae like *Cyclobalanopsis glauca, C. oxydon, C. myrsinifolia*, and *Quercus engleriana*. With the increase of altitude, the dominant species with high specific leaf area, high leaf nitrogen content, and high nutrient content in the community were gradually replaced by species with low specific leaf area, low leaf nitrogen content, and low nutrient content (Salinas et al., 2011). They are deciduous tree species like *Quercus acuteserrata*, *Cyclobalanopsis glauca*, *Betula platyphylla*, and cold temperate coniferous tree species like *Abies fargesii*. The vegetation community in high altitude area is mainly composed of dwarfing dominant species like *Rhododendron hypoglaucum, Sabina Squamata var. fargesii, Fargesia spathacea*, and *Arundinella hirta*. The growth rate of their functional group is low so they adapt to high altitude stress environment.

The animal groups in different vegetation zones along the elevational gradient in the Heritage show the succession process of different functional animal groups with the change of altitudinal gradient in Oriental Deciduous Forest Biogeographic Province. Previous studies have found that animal communities adapt to different environmental conditions along altitudinal gradient mainly by changing their abundance and species turnover (Hodkinson, 2005; McCain, 2009; Sanders et al., 2007). Animals of different functional groups are distributed in different altitudes. Each animal occupies a certain altitude range and coexists with other species in a certain abundance (Hodkinson, 2005). In the vertical vegetation zonation in Shennongjia Natural Heritage, the common animals in evergreen broad-leaved forest are *Macaca mulatta, Crocidura spp., Pica pica*, and *Hirundo rustica*. With the increase of altitude, some animals that adapted to the cold climate of high altitude gradually appeared. For example, the main animal groups in mountain shrub meadow and cold temperate coniferous forest like *Abies fargesii* forest are *Rhinopithecus roxellana, Ochotona huangensis, Carpodacus spp., Parus dichrous*, and *Pyrrhula erythaca* (Li et al., 2003).

The species and quantity of soil fauna and microorganisms in different vegetation zones along the elevational gradient in the Heritage show the bioecological process of soil animals and microorganisms changing with altitude in the Oriental Deciduous Forest Biogeographic Province. Studies have found that the species and quantity of soil animals and microorganisms in the vegetation vertical zonation in mountain also changed along the altitudinal gradient (Bahram et al., 2012). Some specific microbial communities only distribute in a certain altitude range (Gomez-Hemandez et al., 2012). The composition of soil termite community did not change significantly along altitudinal gradient (Palin et al., 2011), but the species change of litter decomposers was more significant with altitude, and the biomass and abundance of soil microbial functional groups also changed significantly along altitudinal gradient (Olson, 1994). Studies have found that the biomass ratio of fungi to bacteria increased with the increase of altitude (Wagai et al., 2011), which was mainly due to the high pressure of high altitude environment, and the soil nutrient cycling of fungi-dominated microbial communities could better adapt to bad environmental conditions (Wardle, 2002).

The vertical vegetation zonation in Shennongjia Natural Heritage site fully represents and shows the integrity of elevational vegetation zoation of in the Oriental Deciduous Forest Biogeographic Province by Udvardy and the bioecological process it breeds. It is a natural experimental site for the vertical differentiation of ecological process in mountain ecosystem in the Oriental Deciduous Forest Biogeographic Province. Therefore, the vertical vegetation zonation has become an important part of the heritage value of Shennongjia Natural Heritage. Further understanding of the bioecological process and its driving mechanism of vertical vegetation zonation in Shennongjia mountain area is an important content in the future research and protection of Shennongjia Natural Heritage. It is also an important problem to solve in scientific management of Shennongjia Natural Heritage.

6.1.3 The Highest Concentration of Global Temperate Genera

Compared with other countries, the concentration of temperate genera in China has the highest in the world. Nearly 931 genera belonging to 78 families are distributed in China. Hubei Shennongjia preserves 590 temperate genera, about 63.4% of the temperate genera in China. The concentration of temperate genera in Shennongjia is higher than that of any other area of China, so it can be concluded that most of the temperate genera in China are distributed in Shennongjia and it is a vital source of global temperate flora. Shennongjia is one of the most important regions for the differentiation, development, and distribution of temperate flora (Figs. 6.4, 6.5, 6.6, and 6.7).

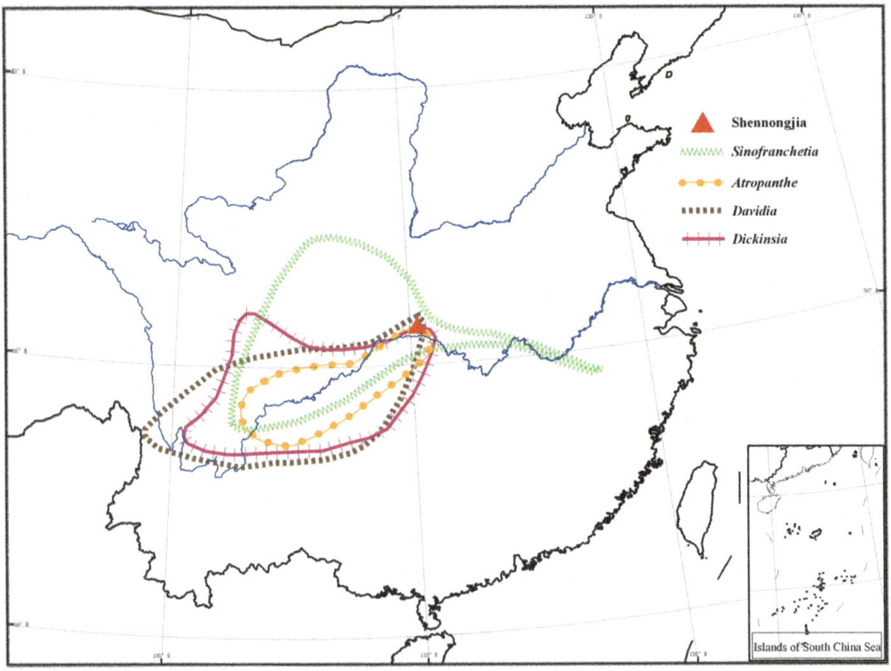

Fig. 6.4 Shennongjia is the disperse source of the single endemic genera to West (according to Ying et al., 1979)

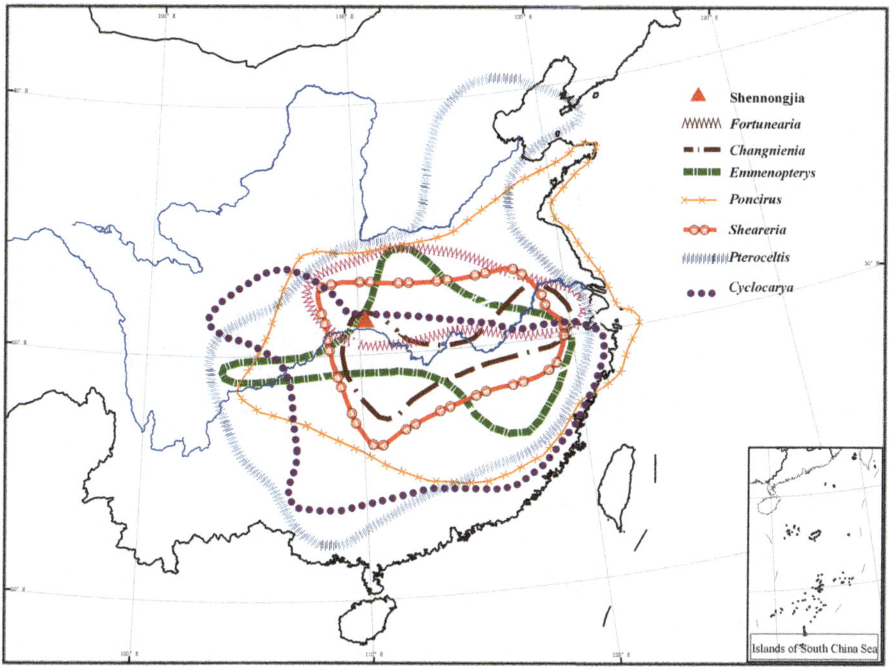

Fig. 6.5 Shennongjia is the disperse source of the single endemic genera to East (according to Ying et al., 1979)

6.1.4 Outstanding Representative of Biological Evolution

With the climate change from hot and humid to warm and humid, the vegetation of Hubei Shennongjia has succeeded into the current forests mixed by evergreen and deciduous broad-leaved trees from tropical and subtropical forests. Towards the end of Devonian times, pollen studies indicate that the forests in Hubei Shennongjia were dominated by Ginkgo and Cycadales flora. During the Period of Cretaceous, the forests dominated by trees from angiosperms and gymnosperms have become widely distributed in Hubei Shennongjia. From the Holocene epoch, the original tropical forest was gradually shifted into temperate forest. Tropical forests dominated by tropical ferns in the Triassic period were gradually replaced by temperate deciduous plants.

Hubei Shennongjia is an intermediate between Sino-Himalayan and Sino-Japanese flora. So it is the center for plants to mix and specialize. Consisting of 590 temperate genera, Hubei Shennongjia is the richest area for deciduous tree species in the world (Fan et al., 2017a). At the same time, Hubei Shennongjia is the most important region of differentiation, development, and distribution for temperate flora. There are many primitive genera and endemic relics concentrated

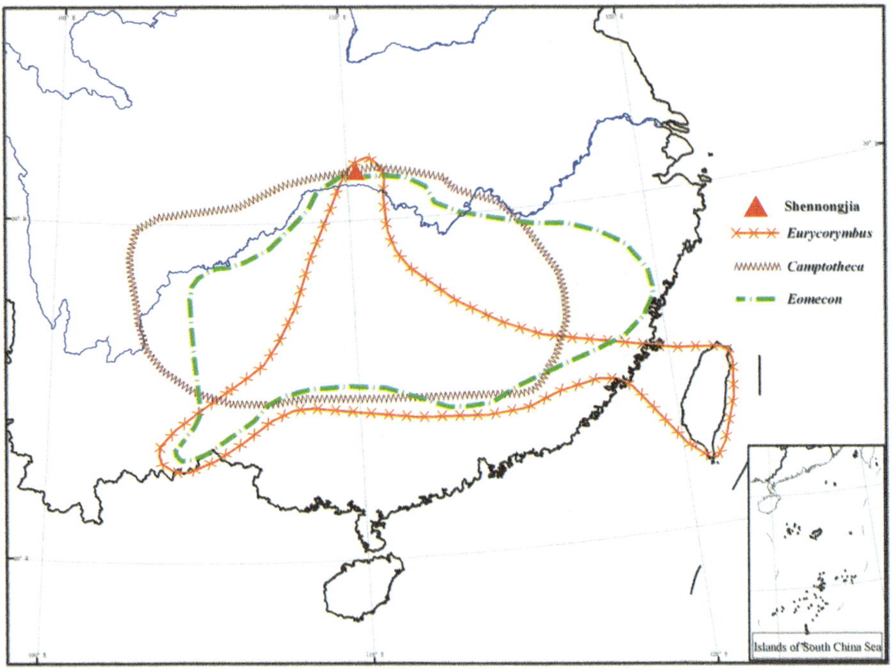

Fig. 6.6 Shennongjia is the disperse source of the single endemic genera to South (according to Ying et al., 1979)

in Hubei Shennongjia, most of them are descendants of the hot-humid flora of the Paleogene period and the Cretaceous period (Fan et al., 2017a).

The evolution processes occurring in Hubei Shennongjia, from hot and humid tropical flora to modern temperate flora, and from broad-leaved evergreen forest to evergreen and deciduous mixed broad-leaved forests, offer a typical model of the evolutionary processes of vegetation communities in Oriental Deciduous Forest biogeographical provinces.

6.2 Biodiversity Value

Hubei Shennongjia contains the richest number of deciduous woody species in the World and are key habitats for numerous relic, rare, endangered, endemic, and type species. So it has outstanding and universal protective and scientific value for its exceptional biodiversity. Hubei Shennongjia is situated in the Qinling-Daba Mountains in Central China. The area's complex terrain has made it less affected by Quaternary Glaciation than Europe and North America. On the other hand, the Qinghai–Tibet plateau uplift led to a unique warm and humid north subtropical monsoon climate at Hubei Shennongjia. This is different from the hot and dry

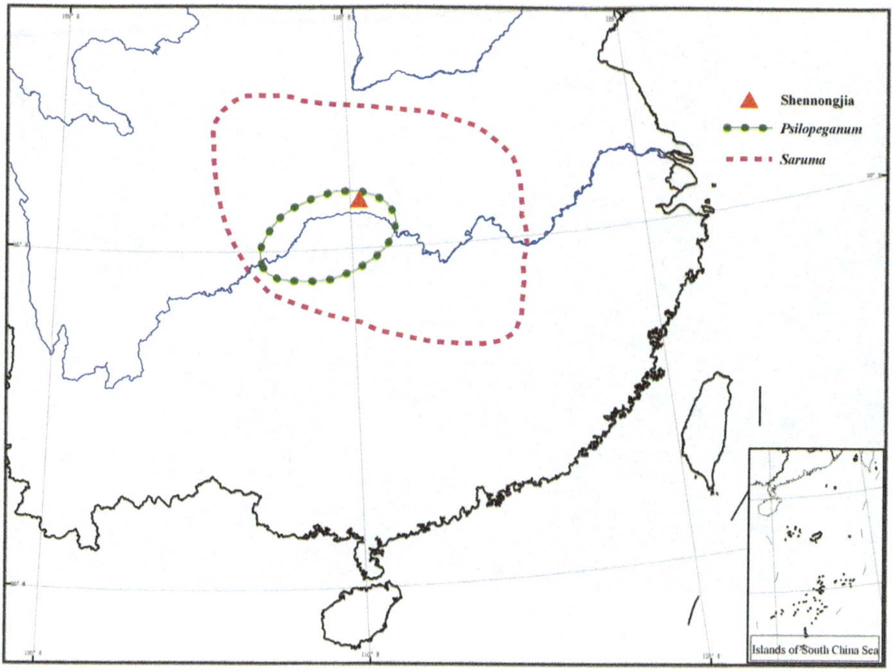

Fig. 6.7 Shennongjia is in the distribution central of single endemic genera (according to Ying et al., 1979)

climates of other areas along the same latitude around the world. This makes Hubei Shennongjia an ideal asylum for Tertiary species from throughout Eurasia.

Hubei Shennongjia contains 3767 higher plants and 629 vertebrate animals, covering 87 mammals, 399 birds, 53 reptiles, 37 amphibians, 53 fish, and 4365 insects. Meanwhile, Hubei Shennongjia possesses major components constituting each class or evolutionary stage of angiosperm, thus providing an integral series of the evolutionary process of angiosperm. In a Chinese context, the area Hubei Shennongjia covers <0.01% of the area of China, but develops 12.5% of Chinese vascular species. It is, therefore, praised as "a green bright pearl" among the subtropical desert and savanna ecosystems along the same latitude globally, containing the richest biodiversity within the Tropic of Cancer zone in the Northern Hemisphere, and thus is a typical model to study the biodiversity and ecosystem processes in subtropical northern hemisphere.

6.2.1 Richful Deciduous Woody Species

Based on evidence from a study by Professor Jun-Sheng Ying, a famous Chinese botanist, we know that Hubei Shennongjia contains the outstanding number of

Photo 6.1 Deciduous broadleaf forests

global deciduous woody species (Ying, 2001). There are 874 deciduous woody plants from 245 genera and 77 families, accounting for 44.5%, 24.8%, and 25.4% of local families, genera, and species, respectively (Photo 6.1).

Worldwide, deciduous woody plants are developed in Europe, the eastern parts of North America, as well as China and Japan. The deciduous plant diversity in North America is greater than in Europe, and Great Smoky Mountains National Park in North America contains the richest number of deciduous woody species. According to the research results, the temperate deciduous forests in the Great Smoky Mountains National Park are regarded as the global richful forests, with the richest biodiversity in terms of deciduous species. It is reported that nearly 200 deciduous woody plants were found in the Great Smoky Mountains, far fewer than the number in Hubei Shennongjia. The temperate deciduous forests mainly distributed in the central part of Japan, where the Ogawa forest is the most representative example. This forest contains the richest number of deciduous species in Japan and is hailed as one of the world's richest temperate deciduous woody areas (Nakashizuka & Matsumoto, 2002).

6.2.2 *Abundant and Intact Ancient/Relic Species*

The evidence of plant and pollen fossil prove that in the Late Devonian period of the Paleozoic era (dating back 350 million years ago), some ancient and primitive taxa such as ferns of Lycopodiaceae, Cycadaceae, and Equisetaceae appeared in Hubei Shennongjia which still persist locally today. The flora of Hubei Shennongjia was basically shaped before the Tertiary period (dating back to 65 million years ago), after a long history of geological events and climatic changes. In the early Tertiary, the northward subduction of the Indian Plate uplifted the Tibetan Plateau rapidly. This gradually made the local climate cool and moist, in contrast to the hot and dry weather of other areas in the Tropic of Cancer. This climate facilitated a large number of species to differentiate, reproduce, and adapt within Hubei Shennongjia. The mild climate and complex topography meant that Hubei Shennongjia, unlike Europe and America, did not experience extensive species' extinction as a result of Quaternary Glaciation. As a result, Hubei Shennongjia has preserved intact ancient/relic species.

Evidence in this regard is:

1. Molecular phylogenetic analysis showing that many local vascular plants from 597 genera and 139 families originated before the Tertiary era, accounting for 55.7% and 65.9% of total genera and families at the site, respectively. This indicates the primitive characteristics of flora in Hubei Shennongjia. The current floristic composition is basically a descendant of the Tertiary flora, and diffuses to adjacent areas in different directions via multiple routes (Ying et al., 1979).

2. Further analysis has revealed that Hubei Shennongjia possesses three out of four families (Ginkgoaceae, Acanthochlamydaceae, Davidiaceae, and Eucommiaceae) known as the most ancient and unique in the modern Chinese mainland. Gymnosperm belongs to more ancient origin taxa than angiosperm in seed plants (Tables 6.2 and 6.3).

Hubei Shennongjia is also rich in the most primitive angiosperm (class Eudicots) families such as Lardizabalaceae, Hamamelidaceae, etc., (Photo 6.2).

Besides, Hubei Shennongjia contains 26 monotypic families (families that only contain one genus and one species) in the flora of China, as many as 42.3% of Chinese flora. The monotypic families are known to be of ancient evolutionary origin (Wu, 2010) (Photo 6.3).

3. Further analysis on the origin of genus has revealed that Hubei Shennongjia contains 22.4% of Chinese endemic genera, in fact, China contains 245 endemic genera (Table 6.4), and most of them (>95%) belong to mono- or oligo-genus, a characteristic of ancient evolutionary origin (Wu, 2010).

Compared with North America, East Asia especially Central China has the highest diversity of endemic genera at the basal gymnosperm and angiosperm taxa (Qian, 2001), reflecting the ancient features and evolutionary integrity of Central China flora. Indeed, Hubei Shennongjia contains as many as 40.7% of the Chinese endemic species in the floristic region of Central China (Qi et al., 1998). Therefore, Hubei Shennongjia is one of the endemic genera distribution centers globally (Photo 6.4).

Table 6.2 Chinese ancient gymnosperm genera, origin time, fossil evidence, and species number in Shennongjia natural heritage

Genus	(MYA) Origin time	Fossil evidence	Species number in World/China/ Shennongjia	Note
Ginkgo	173.3–280.01	The Triassic period	1/1/1	
Abies	130.0–270.02	Eocene	79/25/2	
Keteleeria	130.0–270.02	Late Cretaceous	4/4/1	
Picea	130.0–270.02	Late Cretaceous	52/23/3	
Pinus	130.0–270.02	Early Cretaceous	155/28/5	
Pseudotsuga	130.0–270.02	Oligocene-Pliocene	7/4/1	Disjunct between East Asia and North America
Tsuga	130.0–270.02	Eocene	13/5/4	Disjunct between East Asia and North America
Cunninghamia	168.0–259.03	Early Cretaceous	2/2/1	
Metasequoia	168.0–259.03	Early Cretaceous	1/1/1	A famous relic plant
Cupressus	168.0–259.03	Middle Jurassic	25/6/2	
Juniperus	168.0–259.03	Paleocene	85/22/5	
Platycladus	168.0–259.03	NA	1/1/1	Disjunct between East Asia and North America, endemic genus
Cephalotaxus	138.0–2314	Middle Miocene	13/9/3	
Amentotaxus	138.0–2314	Late Cretaceous	7/4/1	
Taxus	138.0–2314	Middle Jurassic	12/5/2	
Torreya	138.0–2314	Middle Jurassic	8/3/1	Disjunct between East Asia and North America

MYA: Million years ago
NA: No Data

4. From a global perspective, the disjunctive distribution of a family/genus of plants and animals indicates their ancient features. In China, woody species from 49 genera and 34 families are disjunctively-distributed between East Asia and North America, while Hubei Shennongjia contains as many as 76.5% of these genera and 69.4% of these families. The genera including *Liriodendron Pseudotsuga Metasequoia Houttuynia Chimonanthus, Nelumbo Sinofranchetia Dysosma Eomecon Liquidambar Actinidia Clematoclethra Stephanandra Kerria Rhodotypos Tetrapanax Phryma* etc.

Table 6.3 The diversity of the basal taxa of angiosperm families of the Shennongjia Natural Heritage, China

Family	(MYA) Origin time	Number of genera	Number of species	Note
Magnoliaceae	84.5	15/11/5	246/100/14	Liriodendron is disjunct between East Asia and North America
Illiciaceae	61.5	1/1/1	42/28/3	
Schisandraceae	122.5	2/2/2	50/30/7	Schisandra is disjunct between East Asia and North America
Tetracentraceae	61.5	1/1/1	1/1/1	
Eupteleaceae	165.9	1/1/1	2/1/1	
Cercidiphyllaceae	55.8	1/1/1	2/1/1	
Lauraceae	60.4	45/22/9	3000/420/39	Sassafras and Lindera are disjunct between East Asia and North America
Ranunculaceae	71.1	59/41/21	2500/720/101	
Circaeasteraceae	88.9	1/1/1	1/1/1	
Ceratophyllaceae	145.1	1/1/1	6/5/1	
Nelumbonaceae	148.6	1/1/1	2/1/1	
Nymphaeaceae	125.4	5/3/1	60/11/1	
Berberidaceae	71.1	17/11/7	650/320/36	
Lardizabalaceae	88.9	9/6/6	50/40/12	Sinofranchetia, an endemic genus of China, is disjunct between East Asia and North America
Menispermaceae	94.8	71/19/6	450/83/9	
Aristolochiaceae	80.5	7/5/3	475/15/9	Saruma is the primitive genus of the family, originated in the Central China
Piperaceae	80.5	8/4/1	3000/70/1	
Saururaceae	80.5	4/3/2	6/4/2	Disjunct between East Asia and North America
Chloranthaceae	135.8	4/3/1	70/15/4	
Papaveraceae	142.2	23/11/10	200/62/37	
Eucommiaceae	108.0	1/1/1	2/1/1	
Calycanthaceae	120.7	3/2/1	7/4/1	Disjunct distribution from East Asia and North America

This is convincing evidence that Hubei Shennongjia is the distribution center of these species and preserves the most primitive woody species in the country. In terms of animals, some salamanders distributed in Hubei Shennongjia, such as the Wushan salamander (Photo 6.5), the Chinese hynobiid (*Hynobius chinensis*), and Chinese giant salamander (Photo 6.5), show interspecific disjunctive distribution, revealing the impact of major geological and climatic cases, for example, the Himalayan orogeny, the uplift of the Tibetan Plateau and Quaternary Glaciation,

Photo 6.2 Chinese distylium (*Distylium chinense*)

Photo 6.3 Big cone spruce (*Picea neoveitchii*)

on their distribution pattern, and as well as proving their primitive features (Fei & Ye, 1982).

5. The evidence of in situ plant and pollen fossil demonstrates the evolutionary continuum and integrity of primitive plant species in Hubei Shennongjia. Plant and pollen fossils at various geological times, such as *Lycopodium, Selaginella* of the Paleozoic age, *Osmundaceae, Dicranopteris* of the early Mesozoic age,

Table 6.4 Chinese endemic angiosperm monotypic families and some ancient endemic genera of Shennongjia natural heritage

Endemic angiosperm monotypic family	(MYA) Origin time	Some ancient endemic genera	(MYA) Origin time
Bretschneideraceae	41.3	Ginkgo	173.3–280.0
Circaeasteraceae	87.9	Metasequoia	168.0–259.0
Davidiaceae	82–89	Sinofranchetia	44.4
Dipentodontaceae	69.2	Eucommia	108.0
Ellisiophyllaceae	43.3	Dipteronia	41.4
Eucommiaceae	108.0	Chimonanthus	60.4
Ginkgoaceae	173.3	Saruma	40.2
Nandinaceae	35.5	Fortunearia	38.7
Phrymaceae	48.7	Cyclocarya	34.4
Sargentodoxaceae	44.4	Pteroceltis	54.1
Tetracentraceae	61.5	Sinowilsonia	38.7
		Dipelta	35.1
		Sheareria	90.0
		Loxocalyx	43.3
		Triaenophora	43.3
		Davidia	82.1–89.2

Photo 6.4 Dove tree (*Davidia involucrata* var. *vilmoriniana*)

Lygodium, Pinus, Juglans, Ulmus of the Jurassic age, *Taxus, Fagus, Magnolia* of the Cretaceous age, and *Abies, Cyclocarya, Salix, Liriodendron, Sassafras* of the Tertiary age, were all found in Hubei Shennongjia.

Photo 6.5 Wushan salamander

In conclusion, the stunningly rich ancient and relic species, as well as abundant plant and pollen fossils in Hubei Shennongjia, faithfully record the ecological and evolutionary processes of Central China's biodiversity during the past 350 million years. They record how continental drift, geological events, and climate changes have made local species differentiate, become extinct, reproduce, and adapt, providing a living picture of biological evolution.

6.2.3 Critical Habitat for Endangered and Endemic Species

A number of studies show that Hubei Shennongjia is located at the heart of the Central China floristic section of East Asia Subtropical flora. This makes it the gathering place for Chinese rare/endangered/endemic species. Therefore, it is a priority area for biodiversity conservation.

There are eight gathering places for China's rare and endangered species recorded in the IUCN protection list. Hubei Shennongjia contains 234 rare and endangered species from 137 genera and 58 families, accounting for 6.3% of the local vascular plants. Among them, 110 plant species are included in the IUCN Red List (2014), belonging to 83 genera and 47 families. There are 70 species listed as vulnerable (VU), 33 listed as Endangered (EN), and seven listed as Critically Endangered (CR) (Qin et al., 2017; Xie & Shen, 2018). Ninty-four species in Shennongjia are listed in CITES Appendixes I, II (2014) and 167 plant species are included in the China Protected Species List, belonging to 84 genera and 28 families, accounting for 8.6%, 7.1%, and 7.4% of the total n species, genus, and families in protected area (Photos 6.6 and 6.7) (Xie et al., 2017). The number of plant species in Hubei

Photo 6.6 Common bletilla (*Bletilla striata*)

Shennongjia which are included on the IUCN Red List, as well as the protection ratings attributed to them, is significantly higher than those of other Northern Subtropical areas.

In terms of animal species, Hubei Shennongjia contains 130 rare and endangered wild species. Among these, 30 are included on the IUCN Red List (2016), 73 are included in Appendixes I, II of CITES (2014) (Zhou & Wu, 2019). In particular, the site offers ideal habitats for Chinese giant salamander, as well Sichuan snub-nosed monkey, and the Chinese serow.

The property is one of the Chinese key distribution centers of endemic species (Ying, 2001). In Hubei Shennongjia there are 205 local endemic plant species, belonging to 131 genera and 58 families, accounting for 27.5%, 12.2%, and 6.1% of total families, genera, and species, respectively. Hubei Shennongjia also contains 284 local endemic insects, about 6.6% of the total local insect species. The recent discovered two new plant genera, *Zhengyia* and *Repenticaulia*, not only increases the number of Chinese endemic genera, but also fills the gap that Hubei Shennongjia

Photo 6.7 Shiningleaf rose (*Rosa lucidissima*)

had. Previously it had lacked local endemic genera, despite the fact that local endemic species were abundant and local endemism was prominent. The outstanding endemism also reflects the unique environmental conditions which not only favor survival and reproduction, but also promote differentiation and evolution of plants.

Hubei Shennongjia contains 1793 Chinese endemic plant species, belonging to 577 genera and 136 families. It also contains 91 Chinese endemic vertebrates. Hubei Shennongjia offers the habitat for the Hubei subspecies of Sichuan snub-nosed monkey. There are also 55 Chinese endemic plant genera distributed at the site, accounting for 22.4% of the total in China. Such an astonishingly high proportion is unique in the country. Of them, a lot of species are flag or representative species in Subtropical Zone (Photo 6.8).

6.2.4 *Native Habitat of Rare and Horticulture Species*

The property has not only a legend that a figure called Shennong (the Divine Farmer) tested a 100 varieties of herbs in the area and began to use them for traditional Chinese medicine, and also attracted many international naturalists since as long ago as the eighteenth century. Scholars from France, Russia, the United States, Britain, Germany, Sweden, Japan, and other countries visited Hubei Shennongjia more than 14 times in order to collect plants and conduct scientific expeditions. Their discoveries in China (including in Hubei Shennongjia) inspired international enthusiasm for plant collection in the country. Their findings established a milestone of Chinese plant systematic. These discoveries had a far-reaching impact on world botanical studies.

Photo 6.8 Shinningleaf beech (*Fagus lucida*)

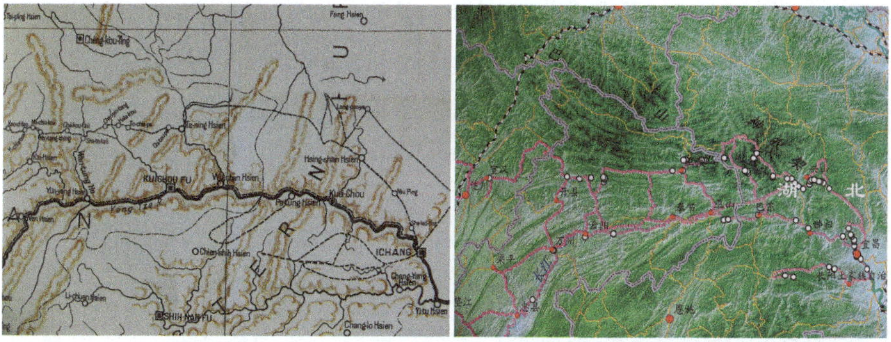

Fig. 6.8 Collection plants by Wilson in Western China including Shennongjia (revised from Ying, 2010)

The type species discovered in Shennongjia have become the standard of studies in the same group globally. From 1884 to 1889, more than 500 new species, including 25 new genera and a new family (Trapellaceae) in Hubei Shennongjia and the surrounding areas were harvested. Based on his experiences in China, Henry published the monograph "*Notes on Economic Botany of China.*" One century later, the Irish youth Seamus O'Brien organized three field trips for bio-surveys in China, following in the footsteps of Henry. He published a book in 2011 named "In the Footsteps of Augustine Henry" in memory of Henry's plant collection in Hubei Shennongjia (Fig. 6.8).

The British botanist Wilson also visited the Pan-Shennongjia region many times to collect plant specimens. Based on these visits, the book "*Plantae Wilsonianae*" was published, recording the 3000 or more species of woody plants he collected in Central and Western China in the years 1907, 1908, and 1910. Today, this book is still an important reference for studying the Chinese woody plants and flora in Hubei and Sichuan provinces. These research monographs offered a priceless window for international scholars to understand China's flora at that time. They created a great sensation in Botany worldwide and had been inspiring the studies of national and international scholars on the flora of China. For example, Professor Shachand from Harvard University published the book "A comparison of Eastern Asiatic and Eastern North American Woody Plants." This book compared woody plants in the northern section of the Grand River of Texas with counterparts in northern parts of the East Asia region (the region with latitude above 22.3 degrees, which includes the Pan-Shennongjia region). Li Hui-Lin, from the University of Pennsylvania, has published a study entitled "Floristic Relationships between Eastern Asia and Eastern North American" (Li, 1952). This monograph, which is a milestone in Floristics, has a scope which involves the Pan-Shennongjia region of Central China (Fig. 6.9).

Through a comprehensive study of the flora of the Pan-Shennongjia region, modern Chinese and international scholars have discovered that the proportion of its endemism is surprisingly high. They have recognized this region as one of the differentiation centers of Chinese endemic plant genera and an area which is of

Fig. 6.9 Plantae Wilsonianae (Revised from Ying 2010)

global significance. Hubei Shennongjia contains 523 type species, accounting for 15.3% of the total species of native vascular plants. Notably, one out of 25 type plant specimens in China comes from Hubei Shennongjia. There are also 320 type insects, accounting for 7.4% of the total native insect species. For these reasons, Hubei Shennongjia is listed as a "critical area of terrestrial biological diversity of global significance" and as an "important type specimen hotspot" by *China Biodiversity*. In the Late nineteenth century, the scientific expeditions by the French missionary Abbe Davids and the British botanist Augustine Henry to the Pan-Shennongjia region made the world realize that the mountains and canyons bore a lot of marvelous flowers and ornamental trees. During the period of Late nineteenth century and early twentieth century, naturalists such as Wilson introduced a large number of native plants to the world. Wilson wrote a book named "China Mother of Gardens" which was a pioneer in promoting the development of horticultural science.

Hubei Shennongjia's unique and stunningly rich native plants have attracted many botanists and horticulturists from around the world. These have introduced a lot of species to Europe and America. For example, in 1980, the Botanical Society of the United States took advantage of Sino-US cooperation to participate in a joint scientific expedition within Hubei Shennongjia. They collected and preserved 402 native plant germplasm resources. Twenty two years later, a follow-up study found that 187 species (46.5%) grew in 18 research institutions worldwide, including those in Denmark, Switzerland, the United Kingdom, Canada, and the United States. This study was widely discussed in the international community. In conclusion, Hubei Shennongjia has made an outstanding contribution to the beautification of the living environment and the development of horticulture science. Hubei Shennongjia has also provided extensive opportunities to carry out studies in areas such as species domestication, species genetic modification and variation within species under different environmental conditions. Highlighting the in situ conservation at the site has a global universal value.

6.2.5 Scientific Shrine for Plant Systematics and Horticultural Science

Hubei Shennongjia sustains abundant and varying flagship and representative species in Subtropical Zone, as well typical forest ecosystem, provides a prominent model to study the biodiversity and ecosystem processes for the Oriental Deciduous Forest biogeographical province. Since the nineteenth century, Chinese scientist and international scholars have conducted systemic studies on geological features, physical geography, biology, and ecology in the site, and have also published more than 620 relating scientific articles or books. To achieve effective protection of the snub-nosed monkeys, the *Conservation and Research Center of the Sichuan Snub-nosed Monkey of Hubei Province* was established in December 2007 within Hubei

Shennongjia. The State Forestry Administration also approved the establishment of the first *State Golden Monkey Research Center*. From population dynamics to molecular genetics, the biology and ecology of the Sichuan Snub-nosed Monkey have been systemically studied. To protect the critically endangered and Chinese giant salamander, the first *Chinese Giant Salamander Research center for Reproductive Biology* was established in Hubei Shennongjia. From the ancient Shennong (the Divine Farmer) to the famous naturalists Ernest Henry Wilson, the outstanding universal protective and scientific values of Hubei Shennongjia have been approved. Now, within Hubei Shennongjia, innumerable biological organisms still present an ongoing reproduce, development, and evolutionary process. Hubei Shennongjia is thus a prominent case to study the biodiversity and ecosystem processes for the Oriental Deciduous Forest Biogeographical Province.

6.3 Integrity of Hubei Shennongjia

6.3.1 Protected Status

Hubei Shennongjia is the National property, is also approved as a National Nature Reserve, a National Park, and a World GeoPark. Therefore, Hubei Shennongjia is protected by National, Provincial or local rules, laws and ordinances.

6.3.2 Border and Scope

Hubei Shennongjia and its buffer zone have clear boundaries. The boundaries mainly follow ridge lines, rivers, and altitude or vegetation distributions. The boundaries accepted the current nature reserve's boundaries. The existing protected areas have been extended to guarantee the integrity of the property's heritage value. The demarcation and staking of the boundary of Hubei Shennongjia has been completed. The boundary of Hubei Shennongjia and its buffer zone has already been set up. Both Hubei Shennongjia and the buffer zone are under strict protection.

6.3.2.1 Principles of Hubei Shennongjia Boundary and Its Buffer Zone

To ensure the integrity of the protection being carried out, and to ensure that Hubei Shennongjia included significant biodiversity and habitats, areas were chosen within an elevation range of 420–3106 m. Boundaries are continuous and contain the most representative native landscapes, including intact altitudinal natural belts and integral ecosystems and all rare and endangered native species. Hubei Shennongjia demarcated the boundaries along with valleys, ridges, rivers, or contours or the nature reserve's boundary. The peripheral natural areas of the boundary are to act as the property's buffers, and the buffer zones cannot contain potential air and water pollution.

6.3.2.2 Description of the Borders and Buffer Zone

Hubei Shennongjia has demarcated the buffer's boundaries. They have been demarcated in consideration of the nature reserve's border. The two components' border has been surveyed. All of Hubei Shennongjia and the buffer zones are under strict protection.

Hubei Shennongjia is divided into two parts, Shennongding-Badong and Laojunshan. Shennongding-Badong' boundary starts from the south point of Jinjiaping (N31°26′10″, E110°15′00″), going north-east to Dujiachayuan, then to Hongjiawan. From this point, it follows the contour line of 1300 m to Wutaishan, and then passes through the 1955.2 m peak (N31°20′22″, E110°21′66″) from west to east. Across Zhaojia river, the boundary arrives at Yangpo, and winds along Wensai River to Songziyuan, then goes to the north along the ridge to Jixinjian (Fig. 6.10). From this point, it connects the boundary of Hubei Shennongjia, and continues along the ridge from the peak of Jixinjian to the outlet of the Zhangbao valley and along the Xiangxi River from south to north, then arrives at the outlet of the Yangjuan River. Along the slope of the mountain west to the village from Muyu and Qingtian, it runs along the ridge to the Yazikou, and then it follows the Yema River to the outlet of the

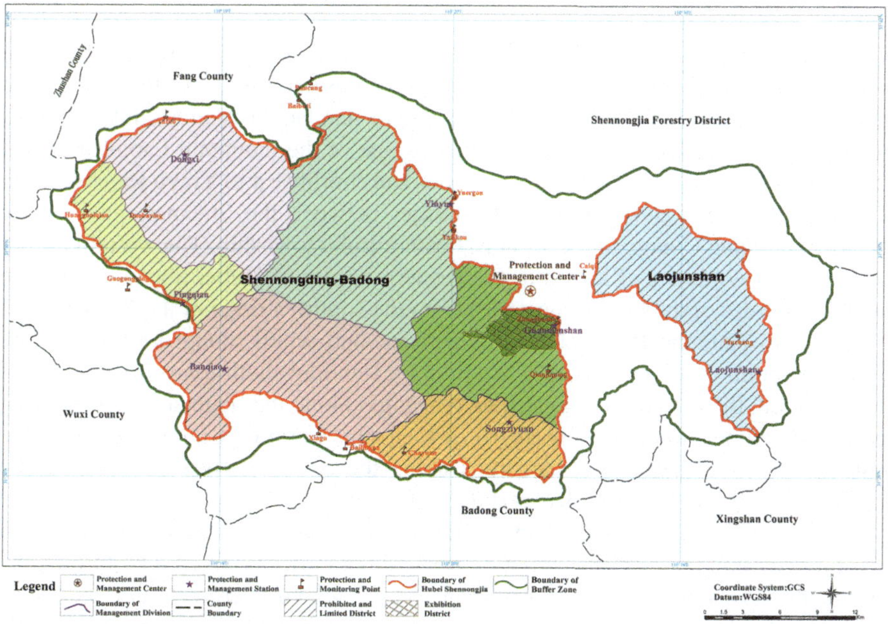

Fig. 6.10 Detailed map of Hubei Shennongjia

Yu'er Valley. From this point, the boundary follows the Yu'er Valley to the peak of Shanziba, and winds along the north-west ridge into the outlet of the Yinyu River (N31°35′21″, E110°13′58″), and then follows the Yinyu River into the outlet of Tianping Valley, and goes along the south-west ridge to the peak (N31°33′32″, E110°13′00″) of the mountain with the elevation of 2471.3 m, and then turns north-west along the ridge to the outlet of the Luoyang River. Across the peak of the next mountain with the elevation of 2046.1 m, the border arrives at the western extremity (N31°31′53.76″, E110°03′22″) of Hubei Shennongjia, west to the village of Huangbaiqian. From this point, the boundary winds to a hillside (N31°26′19″, E110°08′49.44″) between Dajieling and Pingqian, and then from the west part of Xiaojieling, the boundary turns south-east from Xiangsiling and arrives at south-west point (N31°21′38″, E110°09′02″) of Hubei Shennongjia. From the northern ridge of the Dahuangxi Valley, across the outlet of Banqiao Valley, to Xiaguping township, the boundary then returns to the starting point (Fig. 6.10).

Laojunshan, the eastern part of Hubei Shennongjia, the area's border goes from the lowest point (N31°23′23.36″, E110°33′41.23″), southern edge of this part, to Lianghekouzi along the Jiuchong river. After that it runs along the ridge north by east, across the eastern extremity (N31°26′29.81″, E110°33′50.10″) to Taiyangyan, and then goes along the north-west ridge across Sanshiliuguai, Chongping and Wangtianlong until arrives in the peak of the Laojunshan. From this peak, the boundary turns into south-west and runs along the western ridge of Wuguixia valley to Caiqi, and then along the ridge, it winds through the peak of Huanghunling and goes across southern part of the ridge through Majiashan to the southern extremity (N31°21′52.51″, E110°33′17.45″), and runs along the valley and ends on the starting point.

The eastern border of the buffer zone of Hubei Shennongjia runs south by the boundary on the Shennongjia and Xingshan. The northern border of the buffer zone goes northwest from the mountain Motianling to Jiandaogou Valley, and then turns west to Mizigou Valley across the northern slope of the mountain Laojunshan and the peak of the mountain Nanmaojian. It then runs northwest to Yinyu River through Tianciya, Bancang and arrives at the border between the Shennongjia Forestry District and Fang County and ends at Balicao in Jiuhu township. The western border of the buffer zone is 2–3 km from the western border of Hubei Shennongjia. The southern border of the buffer zone goes from the southern ridge of the Xiaohuangxi Valley into Yanduhe River along Yingzuiyan, and winds along the ridge to the altimetric point of 1700 m, then to Wuziping, extending south-eastwards to Tiesuo. At this point, the boundary goes east to Wagangxi along the ridge, and arrives at the peak of Guojian Mountain (Fig. 6.10). It then runs north-east until arriving at the hillside of Muyu Town, and then runs along the Xiangxi River and ends at the crosspoint with the Jiuchong River (Fig. 6.10).

There are 7388 local residents in the buffer zone, who are engaged in beekeeping, eco-tourism, and tea-planting activities. 95% of the buffer is covered by forests, the dominant forests are evergreen broadleaf forests, mixed evergreen and deciduous broadleaf forests, deciduous forests, mixed needle and broadleaf forests, needle forests and some shrubs. Due to the protection of Yanduhe Provincial Nature

Reserve in Badong Country for the boundary of Hubei Shennongjia, which is bounded on the south by Hubei Shennongjia, there is no buffer zone in the boundary adjacent to the Badong County. The buffer zone in north part of Hubei Shennongjia is narrow because the stiff landform and lower level human disturbance. So all of the buffer zone played active role in buffer Hubei Shennongjia against human encroachment. Although some part of the zone is narrow, it can protect Hubei Shennongjia from human disturbance due to the high mountains and low intensity of human activities.

6.3.3 Area of Hubei Shennongjia and Its Attributes

Hubei Shennongjia covers 73,318 ha and its buffer covers 41,536 ha. A small number of people live here, but the site retains a near pristine natural environment. This intact environment holds the richful biodiversity and special ecosystem. Hubei Shennongjia is strictly protected by the center or local government's ordinances. Human activities are prohibited in the buffer area of the property. The area of Shennongjia which has been designated as Hubei Shennongjia contains all of the elements that make the area a place of outstanding value, namely the unique biodiversity, the conservation of numerous habitats, and other features of biological and ecological value.

First, Hubei Shennongjia contains the richful plants and animals and their suitable habitats. It contains the world's most abundant distribution area of deciduous woody plants. Secondly, the north and south slopes of the Shennongjia area contained within Hubei Shennongjia are outstanding examples of elevation vegetation. The property is very important refuge for relic species. These parts of Hubei Shennongjia illustrated the dynamics of plant and animals and their distribution in the global northern subtropical areas. Those attributes of biodiversity and ecosystem processes within Hubei Shennongjia are been under strictly protection.

There are six kinds of first-level (accounting for 46% of total number of first-level habitats globally) and 22 kinds of second-level IUCN/SSC habitats in Hubei Shennongjia. The main habitats include forests, shrublands, grasslands, and wetlands, which serve as a refuge for wild fauna and flora during the Quaternary period. There are 11 types of vegetation and 46 formations in Hubei Shennongjia, hosting exceptional biodiversity. The intact altitudinal vegetation spectrums and various geographic types provide important habitats for biodiversity. For example, the deciduous broad-leaved forest, mixed coniferous and broad-leaved forest, coniferous forest distributed in an elevation range from 1600 to 3000 m provide abundant and seasonal food, as well as complementary space, for the endangered golden snub-nosed monkey of Hubei subspecies. Another example, the understory wet site of the deciduous broad-leaved forest at the north slope of mountains with an elevation of 1700–2200 m in the property, is the ideal habitat for an endangered Orchid, *Changnienia amoena*. The property has a large area of virgin forest without human's trace, is the best protected area for biodiversity and original habitat in the north subtropical zone, providing the highest quality of spaces for the living and reproduction of numerous endangered and endemic species. As such, Hubei Shennongjia can guarantee the integrity of habitat for, as well as the best protection of local biodiversity, and is the most important asylum for representative and relic species of Shennongjia.

The result of flora analysis and fossil evidence strongly prove the integrity of biodiversity and ecosystem in Hubei Shennongjia, mainly due to the absence of invasion of the Quaternary glaciers. As such, Hubei Shennongjia can best represent the succession process of mountain biodiversity and ecosystems in the north subtropical zone. There are various geomorphic types in Hubei Shennongjia, covering karst, fluvial landforms, glacial geomorphology, and tectonic geomorphology. The elevation difference is up to 2700 m, forming suitable climate from the bottom to top, developing mountain yellow soil, mountain yellow brown soil, mountain brown soil, mountain dark brown soil, and mountain podzolized dark brown soil, consequently breeding the special vegetation zonation. Such intact altitudinal vegetation spectrum is the best representative within the Oriental Deciduous Forest Biogeographic Province. The best preserved mixed evergreen and deciduous broad-leaved forest in the north subtropical zone in Hubei Shennongjia, is the essential ligament between the broad-leaved deciduous forest and evergreen forest. Hubei Shennongjia preserves 590 temperate genera, is a vital source of global temperate flora. In conclusion, Hubei Shennongjia can guarantee the most complete protection of essential elements responsible for the ongoing biodiversity processes of mountain ecosystems and communities of plants and animals in the north subtropical zone.

6.3.4 Negative Effects

Currently, the residents living in Hubei Shennongjia and the surrounding buffer zone mainly use the forest for gathering wood and for growing tea. These activities have a certain degree of influence on the environment and resources of the properties, but do not have a large affect. There are currently some tourism in parts of Hubei Shennongjia. At some of the hot spots, the number of tourists is too high during the height of the travel seasons. The Hubei Shennongjia Authority have mitigated disturbing effects by guiding and separating tourists, and by replaced buildings that was not harmonized with the local biodiversity and ecosystem integrity. They formulated regulations on protection, which ensured the property's integrity. Hubei Shennongjia has been protected by harmonizing the local lifestyles and nature. Mining industry, deforestation, and poaching are prohibited in Hubei Shennongjia. Now, tourists became more and more, but it is generally permitted only in the buffer zones.

6.3.5 Protection and Management

Hubei Shennongjia is under the protection by the state and has national or provincial protection status. Hubei Shennongjia is covered by the rules or ordinances issued by the center or local government. Hubei Shennongjia Authority has established three-level management system, including center government, province government, and local government. In addition, a comprehensive system containing the sectors from government, research units and communities has also been set up. Both of these organizations have sufficient finances and staff. Hubei Shennongjia Authority has issued Master Plan for Hubei Shennongjia and Regulation of Shennongjia National Park in Hubei Province for the property.

In view of the existing threats and challenges, Hubei Shennongjia has taken measures to meet the demand of long-term management. Management rules were drawn up in accordance with protection specific demands. The organization and staff ability are strengthened, aiming to comprehensively improve the level of management. The zone management method has been applied in order to optimize the comprehensive value and strict protection of Hubei Shennongjia. Improve the monitoring and early warning systems to minimize the impact of natural disasters on Hubei Shennongjia. The occurred epidemic situation has been controlled timely by the Prevention and Quarantine Station of Forest Diseases and Insect Pests in Hubei province. Tourism management strictly controls the total number of tourists below the limit of the environmental capacity. By restricting influx vehicles, as well as limiting construction of tourism facilities, the air, water, and noise pollution are strictly controlled. The scale of tea farming is controlled in the current level and developed only in the buffer zone.

6.4 Comparison with Other Heritage Site

We had made comparison between Hubei Shennongjia and other properties which are under the same criteria. In the lists from UNESCO, Hubei Shennongjia is one of the mountain properties, so we compared it with mountain properties that have been listed as World Natural Heritages in the same biogeographic province, or other candidate heritages. The site has been compared focusing on biodiversity and ecosystem.

6.4.1 Comparison with Mountain Heritage Sites

Until 2020, 213 natural properties and 39 mixed properties had been listed as World Heritage. According to the regulations, a World Heritage mountain should be with at least 1500 m of elevation difference; the area must have a minimum size of

10,000 ha; the area should fall under an IUCN protected area category I-IV. Until now, 65 World Heritages belong to the mountain heritages. These properties were listed as the world heritages according to criterion vii, criterion viii, criterion ix, and criterion x.

Hubei Shennongjia lies in the subtropic area and in the transition zone from the plains to the plateaus in the eastern Eurasian continent. Consequently, only 13 from the 65 listed mountain properties can be compared with Hubei Shennongjia (Table 6.5). Hubei Shennongjia has a vertical elevation of 2700 m. It has six vertical natural belts, which are (from bottom-up) evergreen broad-leaved forest, evergreen and deciduous broad-leaved mixed forest, deciduous forest, conifer and broad-leaved mixed forest, conifer forest and subalpine shrub/meadow. Hubei Shennongjia has 3767 higher plants, 874 deciduous woody plants, 87 mammals, 53 reptiles, 399 birds, and 4365 insects. Hubei Shennongjia has 234 rare or endangered plant species, and 130 endangered animal species.

(1) Kluane.

This property is located at North 58° which is in the transitional zone from temperate to frigid zones. The main vegetation types are coniferous forest and subalpine tundra. The main advantage of the heritage site lies in its habitats for many animals. The typical belt of the vertical natural belts of vegetation is coniferous forest.

(2) Volcanoes of Kamchatka.

The main feature of this heritage site is its volcanoes. It has more than 800 vascular plants species, and the dominant species are birch, spruce, larch, and other coniferous species, as well as some broad-leaved plant species such as poplar, alder, and willow.

(3) Te Wahipounamu.

The heritage site is located in the western part of New Zealand, the temperate area in the South hemisphere. Its highest altitude is 3764 m. The vegetation is mainly composed of alpine shrubs and also includes temperate rainforest, deciduous forest, subalpine shrub/meadow, and alpine meadow. Seventy-five percent of the heritage is dominated by southern beech and podocarps.

(4) Grand Canyon.

The National Park is known as the Grand Canyon. The heritage site only possesses 1000 vascular plant species, 76 mammals, 299 birds, and 41 reptiles.

(5) Yellowstone Park.

The Park is in the north temperate region, and is known for its geysers. It is famous for its biodiversity. It only has 1100 vascular plants, so the diversity is much lower than that of Hubei Shennongjia.

Table 6.5 The comparison between the mountainous property with Hubei Shennongjia[a]

No	World Heritage Site	Criteria met	Area (ha) and altitude (m)	Geological coordinates	Main features
1	Kluane/Glacier Bay/Tatshenshini-Alsek	(vii)(viii)(ix)(x)	Area:9,839,121 Altitude: 500–5959	58°10′–59°15'N,135°15′–138°40'W	The heritages include active tectonics, volcanoes, glaciers, and the width of natural processes in rivers, from the ocean to some of the peaks in North America. Coastal and marine environments, snow-covered mountains, calving glaciers, deep river canyons, ford-like islands and rich wildlife abound.
2	Volcanoes of Kamchatka	(vii)(viii)(ix)(x)	Area:3,830,200 Altitude:0–4688	56°19′N, 158°30′E	The prominent volcanic sites in the world, with richful and diverse active volcanoes, including volcanic features of the Kamchatka peninsula. The volcanoes and glaciers become a wonderful landscape. The property is dominated by richful biodiversity.
3	Te Wahipounamu-south West New Zealand	(vii)(viii)(ix)(x)	Area:600,000 Altitude:0–3764	166°26′–170°40′E,43°00′–46°30′S	The park, located in south-west part of New Zealand, is famous for its glaciations covering fjords, rocky coasts, towering cliffs, lakes, and waterfalls. 75% of the heritage is dominated by southern beech and podocarps.
4	Grand canyon	(vii)(viii) (ix)(x)	Area:493,077 Altitude:518–2708	35°43′–36°45′N 111°36′–113°56′W	The grand canyon carved from the Colorado River (nearly 1500 meters deep) is the most spectacular canyon in the world. Located in Arizona, across the grand canyon National Park. Its horizontal strata trace back to the geological history of the past two billion years. There are also traces of prehistoric human adaptation to particularly harsh environments.
5	Yellowstone national park	(vii)(viii) (ix)(x)	Area:898,349 Altitude:1610–3462	44°08′–45°07′N 109°10′–111°10′W	The park covers 50% of the global known geothermal features. It also contains the world's largest geyser. Yellowstone was founded in 1872 and is famous for its wildlife such as grizzly bears, wolves, bison, and wolves.

(continued)

Table 6.5 (continued)

No	World Heritage Site	Criteria met	Area (ha) and altitude (m)	Geological coordinates	Main features
6	Great Smoky Mountains	(vii)(viii) (ix)(x)	Area:209,000 Altitude:258–2024	35°26′–35°47′N 83°05′–84°00′W	The park contains more than 1500 vascular plant species. The park is also home to diverse endemic endangered animal species, including the greatest variety of salamanders in the world. It is an refuge habitat for temperate flora
7	Western Caucasus	(ix)(x)	Area:298,903 Altitude:250–3360	43°30′–44°08′N, 39°53′–40°48′E	The site is one of the human untouched mountain. It contains subalpine and alpine pastures only grazed by wild animals, and the site is covered by undisturbed mountain forests. The heritage contains a diverse ecosystems and biodiversity.
8	Cape Floral Region Protected Areas	(ix)(x)	Area:553,000 Altitude:0–2077	32°36′–34°22′S 18°28′–24°50′W	Located at the southwestern tip of South Africa, the site was placed on the world heritage list in 2004. It is one of the most abundant centers of land ecological diversity in the world. The extended heritage sites include national parks, nature reserves, wildlife reserves, national forest parks, and mountain catchment basins. These different terrains add the characteristics of adapting to Mediterranean climate and periodic fire to the local fine hard leaf Persia shrub vegetation and form the unique landscape of cape plant reserve.
9	Tasmanian wilderness	(vii)(viii)(ix)(x) 1993C(iii)(iv)(vi)	Area:1,407,513 altitude:0–1617	41°35′–43°40′S, 145°25′–146°55′E	The parks are located in areas severely affected by glaciers, covered with cliff canyons, covering more than one million hectares, and are one of the only large-scale temperate rainforests in the world. The remains found in limestone caves are evidence that humans lived there more than 20,000 years ago.

10	Uvs Nuur	(ix)(x)	Area:898,064 altitude:759–4116	49°46′–50°40′N, 90°12′–95°38′E	The Ubusu Basin is a closed basin in the northernmost part of Central Asia. It is named Ubsu Lake. Lake Ubusu is a huge shallow saltwater lake, which is an important habitat for migratory birds, waterbirds, and seabirds. A total of 12 protected areas in the region contain major biota in eastern Asia, Europe mainland Europe. The Siberian prairie ecosystem provides habitat for a variety of birds. Many rare animals live in desert areas, while mountain areas are shelters for some of the world's endangered animals.
11	Lake Baikal	(vii)(viii) (ix)(x)	Area:8,800,000 Altitude:1182–2840	51°27′–55°46′N, 103°43′–109°56′E	Lake Baikal, located in southeastern Siberia in the Russian Federation, is the oldest and deepest Lake in the world. It has surface-free freshwater resources. Lake Baikal, known as Galapagos in Russia, has become one of the most diverse and rare freshwater fauna in the world because of its long history and rare human traces. This fauna is of inestimable value to evolutionary science.
12	Three parallel Rivers	(vii)(viii) (ix)(x)	Area:1,700,000 Elevation:760–6740	29°00′–25°30′N 98°15′–100°20′E	Sanjiang parallel flow National Park, located in the mountains of Northwest Yunnan Province, covers 1.7 million hectares and contains eight geographical clusters. The upper reaches of three major rivers in Asia: The upper reaches of the Yangtze River (Jinsha River), the Lancang River (upper Mekong River), and the Nu River (upper Salvin River) from north to south, through 3000 meters deep steep canyons and up to 6000 meters high mountains. Run hundreds of kilometers in parallel without intersecting. It is one of the most biodiversity-rich regions in China and the world.

(continued)

Table 6.5 (continued)

No	World Heritage Site	Criteria met	Area (ha) and altitude (m)	Geological coordinates	Main features
13	Landscapes of Dauria	(ix)(x)	912,624	49°56′N 115°25′E	The site, a common site of Mongolia and the Russian Federation, is a prominent example of the ecological zone of the Dauria grassland extending from eastern Mongolia to Siberia, Russia, and Northeast China. Periodic climate change, with obvious dry and wet periods, leads to a wide diversity of species and ecosystems of global significance. The grasslands and forests, and lakes and wetlands, serve as habitats for rare animal species, such as white cranes, great bustards, gulls and swan geese, and millions of vulnerable, endangered, or threatened migratory birds. It is also an important location on the cross-border migration path of Mongolian gazelles.

[a]Data source:http://whc.unesco.org/en/list

(6) Great Smoky Mountains Park.

The highest elevation of the Great Smoky Mountains National Park is 2024 m. The upper part of the mountain is coniferous forest which is dominated by Canadian fir and spruce, while the lower part of the mountain is broad-leaved forest composed of tall oaks, pines, and hemlock. In the absence of evergreen constructive species, it lacks evergreen broad-leaved forest and evergreen and deciduous broad-leaved mixed forest, both of which are present in Hubei Shennongjia. There are 1500 vascular plant species, 120 rare or endangered plant species, and 200 kinds of birds. There are more than 1300 native vascular plant, 230 woody plants, and there are 120 rare or endangered plant species, and 200 bird species.

(7) Western Caucasus.

Western Caucasus is located in north temperate area, and the basal belt of the vertical natural belts is deciduous forest. There are 1580 vascular plant species, 384 vertebrates, 60 mammals, and 2500 insect species. About 1/3 of the mountain species and about 1/5 of the tress species are endemic.

(8) Cape Protected areas

The Cape protected areas is located in the temperate climatic area in the South hemisphere, and the highest elevation is 2077 m, and it has 8996 vascular plant species. The flora in Cape is the fynbos, fine-leaved vegetation in both the Mediterranean climate and periodic fires. Evergreen forest is only scattered in the region. However, it lacks evergreen and deciduous broad-leaved mixed forest.

(9) Tasmanian wilderness.

The Tasmanian wilderness is the temperate region in the South hemisphere. It is one of the few large-scale temperate rainforests in the world. It has 1890 vascular plants. The fauna is of global importance due to its unusually high proportion of endemic species and relict groups of ancient lineage.

(10) Uvs Nuur Basin.

Uvs Nuur is in temperate grassland climate zones and mainly includes grassland, lake, and desert ecosystems.

(11) Lake Baikal.

Lake Baikal is mainly composed of lake ecosystems, and the species diversity is mainly that of freshwater animals.

(12) Three parallel rivers.

The Protected areas belongs to the biogeographic province of Mixed mountain and highland systems. It has 6000 plant species, 173 mammals, and 417 birds. Dominant vegetation is evergreen broad-leaved forest type. It belongs to western subtropical area, and its climate is mainly affected by Indian Ocean air current; while evergreen and deciduous broad-leaved mixed forest is the dominant vegetation type in Shennongjia property, and it belongs to eastern evergreen broad-leaved region, and its climate is mainly affected by Pacific air current.

(13) Dauria.

The site, a common site of Mongolia and the Russian Federation, is a prominent example of the ecological zone of the Dauria grassland extending from eastern Mongolia to Siberia, Russia, and northeast China. Periodic climate change, with obvious dry and wet periods, leads to a wide diversity of species and ecosystems of global significance. The grasslands and forests, and lakes and wetlands, serve as habitats for rare animal species, such as white cranes, Great Bustards, gulls and swan geese, and millions of vulnerable, endangered, or threatened migratory birds. It is also an important location on the cross-border migration path of Mongolian gazelles.

Comparison Conclusion

According to the comparison, Hubei Shennongjia is special and different from other 13 sites in terms of ecosystem types, biodiversity, flora and fauna of biogeography, and vertical natural belts. The ecosystem type of Hubei Shennongjia is different from that of Uvs Nuur Basin and Lake Baikal, and the vegetation type of Hubei Shennongjia is different from that of Three Parallel Rivers Areas. Hubei Shennongjia has a greater diversity of species than these sites (namely the Volcanoes of Kamchatka, the Grand Canyon National Park, the Great Smoky Mountains National Park, the Western Caucasus, the Tasmanian Wilderness, and Lake Baikal). It possesses more complete vertical natural belts than the others, for example, Kluane, Te Wahipounamu, Yellowstone National Park, and Cape Floral Region Protected Areas. These vertical natural belts are something which cannot be replicated in other World Heritage mountain sites.

6.4.2 Comparison with Heritage Sites in Same Biogeographic Province

Based on Biogeographic provinces of the world by Udvardy (1975), different biogeographic provinces have different ecological processes and biodiversity. As such, Hubei Shennongjia is not comparable with World Heritage Sites in different biogeographic provinces. Hubei Shennongjia is located in the deciduous forest biogeographic province. There are nine World Heritage Sites in this province (Table 6.6). Among them, six of the sites have been classified as the World Heritage sites according to natural beauty or geological criteria. These criteria are different from the criteria x and criteria ix under which Hubei Shennongjia is being listed. This means that these six sites are not comparable with Hubei Shennongjia.

Mount Huangshan was under Criterion ii (culture), Criterion vii (natural beauty), and Criteria x. The climate of Huangshan is the humid subtropical monsoon climate, and its bedrock is granite. Its elevation is 1864 m, and the elevational vegetation includes evergreen broad-leaved secondary forest, evergreen and deciduous broad-

Table 6.6 Nine world heritage sites which are located in the same deciduous forest biogeographic province with Hubei Shennongjia[a]

No	World heritage site	Criteria met	Important ecological and biological characteristics (Criteria ix)	Biodiversity and habitats (Criteria x)
1	Mount Sanqingshan National Park	(vii)	NA	NA
2	Jiuzhaigou Valley scenic and historic interest area	(vii)	NA	NA
3	Wulingyuan scenic and historic interest area	(vii)	NA	NA
4	Huanglong scenic and historic interest area	(vii)	NA	NA
5	China Danxia	(vii)(viii)	NA	NA
6	Mount Taishan	(i)(ii)(iii)(iv)(v)(vi)(vii)	NA	NA
7	Mount Huangshan	(ii)(vii)(x)	NA	Huangshan is known as the first strange mountain in Sinian country. In the heyday of Chinese history, it was widely praised through literary and artistic forms, such as the mountain and water style in the middle of the sixteenth century. Huangshan is famous today for its magnificent scenery-strange pine growing on granite rocks and strange stones emerging in the sea of clouds. For tourists, poets, painters, and photographers who come to this scenic spot from all sides, Huangshan has eternal charm.
8	Shirakami-Sanchi	(ix)	Located in the mountains of the northern state, the white god mountain is inaccessible, retaining the remains of the last undeveloped beech forest in the cold zone. The beech	NA

(continued)

Table 6.6 (continued)

No	World heritage site	Criteria met	Important ecological and biological characteristics (Criteria ix)	Biodiversity and habitats (Criteria x)
			tree was once widely distributed, covering almost all hills and hillsides in northern Japan. Black bears, iguanas, and 87 species of birds also live in the white god mountain forest.	
9	Fanjingshan	(x)	NA	Fanjingshan was under natural habitats for biological diversity criterion (criterion x). It is covered by diverse types of vegetation, and home to biodiversity. Hubei Shennongjia has lots of endemic species. Fanjingshan is also home to primeval beech forest in the subtropical region.

[a]Data source: http://whc.unesco.org/en/list

leaved mixed forest, deciduous forest, and coniferous forest. There are 1102 vascular plant species, and 19 Huangshan endemic species, 300 vertebrates. Shirakami-Sanchi was under ecological and biological processes criterion (Criterion ix). Its area is 16,971 ha. This site contains the intact cool-temperate forest dominated by Siebold's beech.

Fanjingshan was under natural habitats for biological diversity criterion (Criterion x). It is covered by diverse types of vegetation, and home to biodiversity. Hubei Shennongjia has lots of endemic species. Fanjingshan is also home to primeval beech forest in the subtropical region. Hubei Shennongjia situated in humid subtropical monsoon climate zone has a vertical elevation of 2700 m. Hubei Shennongjia has six vertical natural belts, which are (from bottom-up) evergreen broad-leaved forest, evergreen and deciduous broad-leaved mixed forest, deciduous forest, conifer and broad-leaved mixed forest, conifer forest, and subalpine shrub/meadow. Hubei Shennongjia has 3767 higher plants, 874 deciduous woody plants, 87 mammals, 53 reptiles, and 4365 insects. Hubei Shennongjia has 234 rare or endangered plant species, and 130 rare or endangered animal species.

Comparison Conclusion

Compared with Mount Huangshan and Shirakami-Sanchi in the same biogeographic province, Hubei Shennongjia is special and different in terms of climate zone, zonal vegetation type, biodiversity, and vertical natural belts. Hubei Shennongjia is located in a different climate zone from Shirakami-Sanchi. The zonal vegetation in Hubei Shennongjia is different from Mount Huangshan and Shirakami-Sanchi. The biodiversity in Hubei Shennongjia is abundant. It also owns complete vertical natural belts. In conclusion, Hubei Shennongjia cannot be replicated by other World Heritage in the same biogeographic province.

6.4.3 Comparison with Nominated Properties

There are six properties on the Tentative List of World Natural Heritage in the deciduous forest biogeographic province (Table 6.7). The value of the Yangtze Gorges Scenic Spot mainly focuses on its cultural and esthetic aspects. It has 2093 vascular plant species and 570 vertebrates. The value of Yandang Mountain mainly focuses on its cultural, esthetic, and geological aspects. It has 1248 seed plant species. Mt. Soraksan Nature Reserve was under the criteria of esthetics, geology, and habitats, and its altitude is 1708 m. The vegetation here is dominated by mixed coniferous broad-leaved forest and it has 822 vascular species. Mt. Kumgang and the Historical Relics, whose elevation is 1638 m, is remarkable for its culture, natural beauty, and habitats. The vegetation is mainly composed of mixed coniferous broad-leaved forest, and it possesses 2256 vascular species. Poyang Lake is mainly plain wetland forming a habitat for birds. It mainly protects birds and aquatic flora and fauna. The Alligator Sinensis is the key global habitats for alligator family. The outstanding universal values of Hubei Shennongjia lie in its biodiversity and ecological and biological processes. There are 3767 higher plants, 629 vertebrates, and 4365 insects. The vertical natural belts with the bottom-up order of evergreen broad-leaved forest, evergreen and deciduous broad-leaved mixed forest, deciduous forest, conifer and broad-leaved mixed forest, conifer forest, and subalpine shrub/meadow.

Comparison Conclusion

Hubei Shennongjia is different from the six candidate properties, in terms of inscribed criteria, major ecosystem type, zonal vegetation type, biodiversity, vertical natural belt, as well as protected object. The outstanding universal values in Hubei Shennongjia are different from those of the Yangtze Gorges Scenic Spot and Yandang Mountain. With regard to the reprehensive ecosystem type, Hubei Shennongjia is dominated by forest ecosystem, while Poyang Lake and the Alligator Sinensis Nature Reserve are dominated by wetland ecosystem. Hubei Shennongjia is also different from Mt. Soraksan Nature Reserve and Mt. Kumgang and the Historical Relics in zonal vegetation type. Hubei Shennongjia has stunningly rich biodiversity and complete vertical natural belts. Protecting biodiversity and ecosystem is the major object in Hubei Shennongjia, while the major object of protection in Poyang Lake is birds and their habitat, and the major object of the Alligator site is to protect the Alligator and its habitat. Hubei Shennongjia cannot be replaced by other candidate properties in the same biogeographic province.

Table 6.7 The six properties on the tentative list of world natural heritage in the deciduous forest biogeographic province[a]

No	World heritage site	Criteria met	Area:(ha) Altitude(m)	Geological coordinates	Main features
1	Poyang Lake	(x)	Area: 384,100 ha	115°–116°44′ E/28°25′–29°45′N	Poyang Lake is one of the most famous bird migration sites in the world. There are the whitest cranes in the world, 150 species of birds, many of the world's rare bird species. Twenty species of rare birds inhabit frequently.
2	The Alligator Sinensis nature reserve	(x)	Area: 443,00 ha	180°–119°6′E/30°6′–31°6′N	Alligator is a unique Chinese crocodile, the oldest existing reptile, the world's only two species of alligator family, a small number. Alligator nature reserve provides a suitable living environment for alligator. Currently, more than 1% thousands of alligators live here.
3	Yangtze gorges scenic spot	(vii)(vi)	Area: 120,800 ha	109°34′–110°12′E 30°45′–31°2′N	It is famous for its steep canyon, changing water, and rich and long history and culture. At the same time, there is a biodiversity, including vascular plant from 166 families 2093 species, 570 species of vertebrates, 69 species of animals, 124 species of birds, etc.
4	Yandang Mountain	(viii) (vii)(vi)	Area: 18,600 ha; Altitude 1056 m	121°00′–121°30′E /28°35′–28°55′ N;120°35′–121°59′E /28°10′–28°34′N	Yandang Mountain is a typical model of cretaceous volcanic rocks and a natural museum of rhyolite volcanic rocks. The global mutation event is unique in time and space. It is universal and representative in the giant volcanic (rock) zone on the mainland edge of Asia in the western Pacific Ocean. It is a natural deep drill to study the deep geological process of magmatism on the mainland edge. It contains 1248 seed plants belonging to 160 families. Nanxi River is a tree-like water system with a total length of 145 km, flowing through the east of Yandang Mountain, showing a typical valley landscape. There are Taiwan Shuqinggang, Ginkgo biloba, West China maple poplar, and other national key protection of precious tree species.

5	Mt. Kumgang and the historical relics	(vii) (vi) (x)	Altitude: 1638 m	126°04′E/38°37′N	The relics are at the border of Korea and South Korea, with the peaks stand steeply. According to legend, there are 12,000 peaks, strange rocks, with all things the most spectacular. The waterfalls fly and the forests are gloomy. Preserved temples, stone towers, and many other historical relics. The main vegetation is coniferous broad-leaved mixed forest. It contains 2256 vascular plants, 38 animals, 130 birds, nine reptiles, ten amphibians, and 30 fish species.
6	Mt. Soraksan nature reserve	(vii)(x) (viii)	Area: 16,360 ha; altitude: 1708 m	128°18′06″-128°30′43″ E/38°05′25-″-38°12′36″ N	It has spectacular ridge hills, marveled by geographical features, and retains religious historical and cultural relics such as temples. There are 822 vascular plants, 50 species of birds and mammals, especially some endangered species and rare fish.

[a]Data: http://whc.unesco.org

6.4.4 Summary of the Comparisons

Hubei Shennongjia is unique in its biodiversity, flora of biogeography, elevational vegetation zone, and ecological and biological processes. Based on the comparison, the unique characteristics of Hubei Shennongjia from a global perspective are as follows: It possesses significant biodiversity and a large number of ancient, endangered, endemic and type species, and contains the richest deciduous woody plants in the world; therefore it is of outstanding universal value for conservation and science. It contains the typical representative of mixed evergreen and deciduous broad-leaved forest in the northern hemisphere, and owns typical elevational vegetation zones in the deciduous forest biogeographic province. It is a prominent case on study on ecosystem processes and the mechanism driving altitudinal patterns. It is the cradle of the global temperate flora, and the representative example of the zonal vegetation and succession of temperate taxa in deciduous forest biogeographic provinces.

Chapter 7
Factors Affecting Hubei Shennongjia

At present, Hubei Shennongjia is occasionally threatened by natural disasters and human disturbances. Natural disasters include geological disaster, frozen disaster, forest fire, forest toxic organisms, as well as ecological disasters due to global warming. Human activities include tourism and tea farming. The growing number of tourists has brought challenges to the protection of natural resources and ecological environment of Hubei Shennongjia. Tourism threatens to a certain extent the protection of the wild animal's activities and habitat in Hubei Shennongjia. Influx of excessive tourists and vehicles and construction of reception facilities deteriorate to a certain degree the quality of air, water, and induce noise pollution. Tea farming is the traditional industry in Hubei Shennongjia, is one of the livelihoods of local residents, and has a potential influence on the protection of Hubei Shennongjia.

7.1 Present Conservation State

7.1.1 Protection under National Regulations and Laws

Hubei Shennongjia has been successively protected by many names from center government, provincial government, or local government.

7.1.2 Management System

Administrative authorities have been built for Hubei Shennongjia. These have been provided with adequate staffing and finances for the property. The top administration was from the center government, then the province, then Shennongjia Forestry District, and finally the prefecture.

Z. Xie, G. Shen, *The Outstanding Universal Value and Conservation of Hubei Shennongjia*, https://doi.org/10.1007/978-981-16-0684-7_7

7.1.3 Boundaries and Monitoring System

The administration bureau had issued a master plan for management, demarcated the border, and established a monitoring system. The issued Master Plan of Hubei Shennongjia contains the protective technical documents.

7.1.4 Current Status of Species, Communities, and Ecosystems

Hubei Shennongjia contains 3767 higher plants. It has 629 species of vertebrate animals, including 87 mammals, 399 birds, 53 reptiles, 37 amphibians, 53 fish, and 4365 insects. There are 874 species of deciduous woody plants, and 590 temperate genera. Hubei Shennongjia contains 115 endangered vascular species and 48 animal species on the IUCN red list. Within Hubei Shennongjia, snub-nosed monkeys are found in subtropical evergreen forests, mixed forests, broadleaf and conifer forests at elevations ranging from 1600 to 3000 m above sea level. According to recent surveys, there are about eight groups with a total of 1550 individuals. At the sites such as Jiuchong River and Majia Valley, etc., populations of endangered plants like Chinese yew and Badong manglietia are distributed in a concentrated pattern. There are six kinds of first-level and 22 kinds of second-level IUCN/SSC habitats in Hubei Shennongjia. The main habitats include forests, shrublands, grasslands, and wetlands. There are 11 types of vegetation and 46 formations in Hubei Shennongjia. The mountain soil type's serial of Hubei Shennongjia develops completely, and the typical elevational vegetation zones are the best representative within the Oriental Deciduous Forest Biogeographic Province (Fan et al., 2017a). In Hubei Shennongjia, a large area of land at the sites of YinYu River, etc., is rarely explored by human activities, and the virginal forest has an area of 14,947 ha, well preserving various ecosystems in the north subtropical zone. Hubei Shennongjia, which has just a small amount of human activity, has better maintained its original natural state.

7.2 Factors Affecting Hubei Shennongjia's Ecosystem and Biodiversity

From the 1990s, tourists and the tea plants have been developed in the buffer zone of the property, which caused potential disturbance to Hubei Shennongjia's natural biodiversity and ecosystem elements. However, migration, limitations on the amount of tourists, natural forest protection projects, the Grain for Green project, and subsidies for conservation have all helped mitigate the impact. This situation has shown a significant improvement. Transportation is the foundation infrastructure development and is a prerequisite for Shennongjia to develop its economy and tourism. A network of roads forms the main type of transportation in Shennongjia. National road 209 crosses Shennongjia from the south to the north connecting the main parts of the area. These roads make an important contribution to the area's infrastructure. Shennongjia Airport was completed in May of 2014. The airport is located at a distance of 11 km from the boundary of property. The airport in fact does not cause any influence on Hubei Shennongjia.

7.2.1 Driving Factors for Forest Area

During the 30 years since the establishment of Shennongjia Nature Reserve, the forest area increased from 374.02 km^2 (1980) to 502 km^2 (2015) with an increase rate of 34.27% (Fig. 7.1), and the forest coverage increased from 47.94% to 64.36% with an increase rate of 16.42%. Through comparing the data before (1980) and 10 years after (1990) the establishment of the reserve, it can be seen that the forest area increased from 374.02 km^2 (1980) to 428.45 km^2 (1990) with an increase rate of 14.55% (Fig. 7.1), and the forest coverage increased from 47.94% (1980) to 54.91% (1990) with an increase rate of 6.97% (Wang et al., 2017). Based on the comparison of the data before (1990) and after (2010) the Natural Forest Conservation Program (NFCP) and Sloping Land Conversion Program (SLCP), the forest coverage expanded from 428.45 km^2 to 528.04 km^2 in this reserve with an increase rate of 23.24% (Fig. 7.1), and the forest coverage increased by 12.77% (Fig. 7.1) (Wang et al., 2017). After the establishment of Shennongjia Nature Reserve, the forest

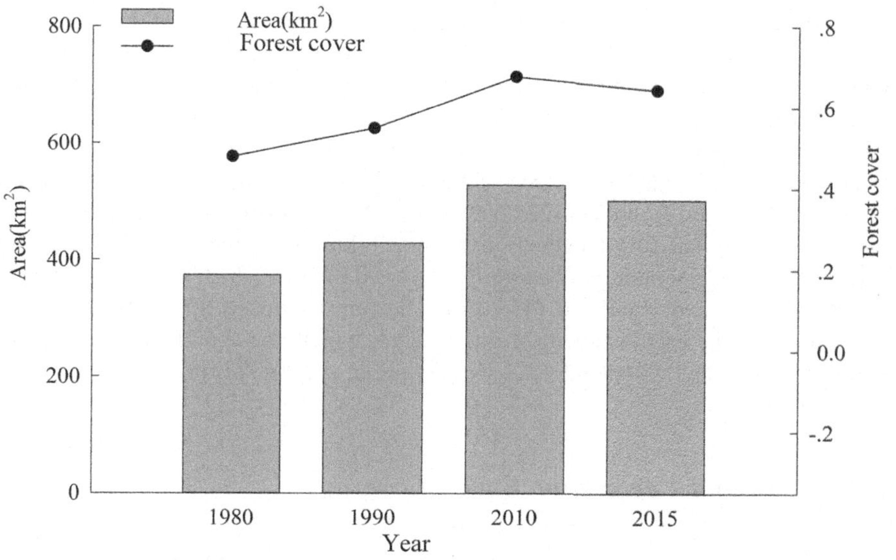

Fig. 7.1 Change of the forests coverage in Shennongjia

coverage expanded significantly during the 30 years, which indicated this reserve was efficient in conserving the forests.

since 1970, the forest industry enterprises in the forest region provide timber products for the state and mainly engage in harvesting timbers, which has caused the excessive deforestation, sharp decrease in the forest growing stock, and destroyed natural vegetation; especially, the habitat and the reproduction of rare animals and plants were limited. The establishment of Shennongjia Nature Reserve in 1982 promoted the change from harvesting to conserving the forest resources and wildlife habitat. In 1986, the State Council approved Shennongjia Nature Reserve as a national forest and wildlife nature reserve (Wang et al., 2017). In 1987, the *Regulations on the Conservation of Natural Re-sources in Shennongjia, Hubei Province* was issued, and the annual output of timber in the forest region reduced by 48,000 m^3. In 1990, approved by UNESCO, Shennongjia Nature Reserve joined the International "Human and Biosphere" Reserve Network, which led to the annual increase in forest growing stock of 170,000 m^3 in this reserve, and the forest was effectively protected.

Shennongjia Nature Reserve achieved good results in mitigating forest loss and increasing forest coverage during the 30 years (Deng et al., 2018). After the establishment of the reserve, the forest area has an increase of 34.27%, the forest coverage has an increase of 16.42%, and 82.77% of the gained forest was restored from shrub and woodland. Wang et al. (2015) suggested that the net reduction rate of forest in the protected areas was 2.07% lower than the national net reduction rate through investigating the changes of forest in 407 national nature reserves in China from 2000 to 2013. Joppa and Pfaff (2010) assessed the forest loss rate in nature reserves of 147 countries around the world, showing that the natural reserves of 109 countries achieved significant results in mitigating forest loss. During the past 30 years (1980–2015), the area of newly gained forest reached 208.67 km^2, 82.77% of which was recovered from shrub (50.85%) and woodland (31.92%). Through comparing the data before (1980) and 10 years after (1990) the establishment of this reserve, the area of newly gained forest was 180.41 km^2, 90.65% of which was recovered from shrub (69.32%) and woodland (21.33%). After the implementation of NFCP and SLCP (2010), 84.78% of the newly increased forest was restored from shrub (62.72%) and woodland (22.06%) (Fig. 7.2). Meanwhile, it was found that the area of the original forest decreased by 80.21 km^2 over the 30 years after the establishment of Shennongjia Nature Reserve, 90.01% of which were converted to shrub (41.31%) and woodland (48.70%). The forest reduced by 33.49% over the 10 years after the establishment of this reserve, which was mainly degenerated to shrub (62.21%) and woodland (26.63%) (Wang et al., 2017) (Fig. 7.2).

7.2.2 Driving Factors for Forest Coverage

In order to explore protecting efficiency of the National Nature Reserve, a dimidiate model is used to measure the fraction vegetation cover (FVC) over the Shennongjia

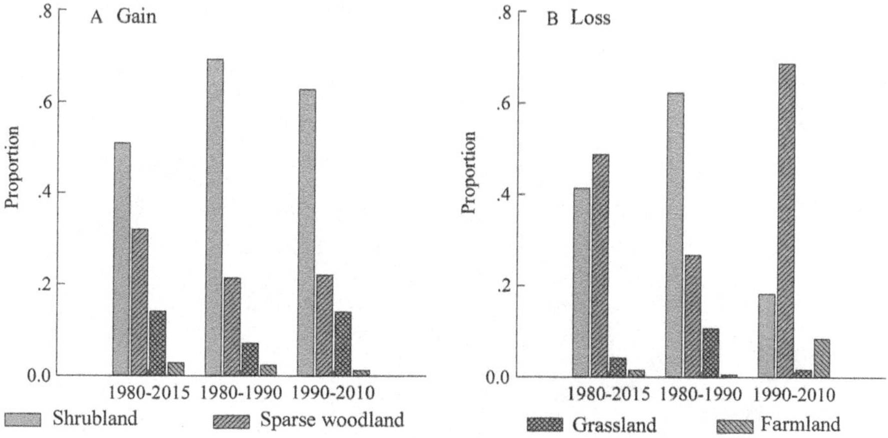

Fig. 7.2 Factors driving forest increase **(a)** and decrease **(b)** in Shennongjia

Forest District from 1998 to 2013 by using the 1 km-resolution, ten-day NDVI serial data of Spot VEGETATION.

7.2.2.1 The Average Forest Coverage of Shennongjia

Between 1998 and 2013, the FVC of Shennongjia Forest District was 66.8% in average, ranging from 64.3% to 71.4%, with the lowest mean FVC in 2012. The maximum FVC of Shennongjia Forest District was 93.8% in average, ranging from 92.5% to 94.5%, with the lowest maximum FVC in 2006 (Liu et al., 2018). The mean FVC inside the National Nature Reserve was 66.5%, outside the National Nature Reserve was 66.8%. The maximum FVC inside the National Nature Reserve was 94.3%, outside the National Nature Reserve was 93.7% (paired T test, $T = 3.221$, $P = 0.006$). The FVC inside the National Nature Reserve was significantly higher than outside the National Nature Reserve. From 1998 to 2013, there were differences in the mean FVC pattern of Shennongjia Forest District (Fig. 7.3). The maximum FVC was 82.4%, and the minimum was 27.7% (Liu et al., 2018). Spatial clustering analysis shows that the higher FVC areas are mainly located in Yinyuhe, Jiuchonghe, Lichahe, Wenshuiliu, and Songluohe, while the lower FVC areas are mainly around Dajiuhu, Songbai, Laojunshan, and the tourist roads inside the National Nature Reserve.

The FVC altitude pattern shows that the FVC is highest in the area between 1500 and 2000 m ($R^2 = 0.185$, $P < 0.0001$). Below 1700 m, as the altitude increases, FVC increases. Above 1700 m, FVC decreases with the increases of altitude (Fig. 7.4) (Liu et al., 2018). The number of arbor species and total species analyses of Zhao et al. (2005) showed that FVC has a significantly positive correlation with both number of arbor species and total species (Fig. 7.4). Both FVC and species richness are highest between 1500 and 2000 m, which is a typical zonal vegetation

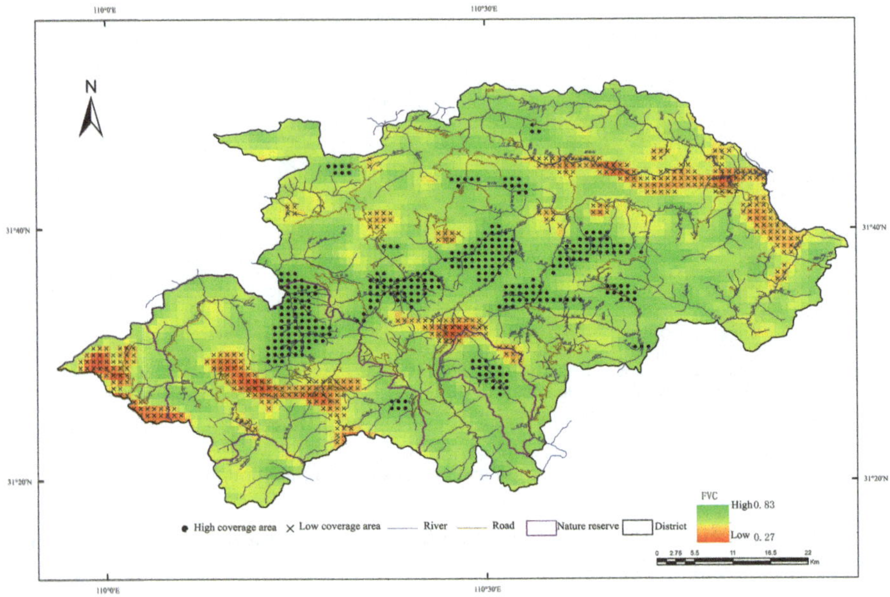

Fig. 7.3 The mean FVC pattern of Shennongjia Forest District

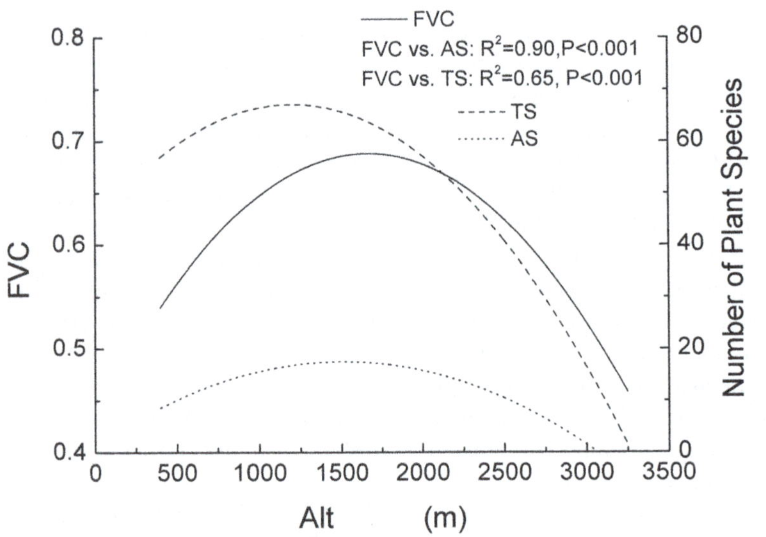

Fig. 7.4 The mean FVC altitude pattern of Shennongjia Forest District and relationship with species richness

distribution area of evergreen and deciduous broad-leaved mixed forest. It further shows that the typical mixed-leaved forest in the Shennongjia Forest District is a relatively well-preserved evergreen and deciduous broad-leaved mixed forest with extremely high coverage and diversity.

7.2.2.2 The Forest Coverage Change of Shennongjia during 1998–2013

The FVC change rate of Shennongjia Forest District during 1998–2013 was 1.45%, with 2.26% inside the National Nature Reserve and 1.23% outside the National Nature Reserve (Liu et al., 2018). The FVC of Shennongjia Forest District has an increase trend during the 16 years. The increase of FVC inside National Nature Reserve was far more than outside the reserve, which indicated that the Shennongjia National Nature Reserve has a good protect efficiency. There has been a significantly increase of FVC in the top of Laojunshan, Shennongding, Qingshu village, Dongxi village, and Xiangsiling village at the southwest of the National Nature Reserve (Liu et al., 2018). There was also a small increase of FVC on both sides of the tourist roads. The areas with reduced FVC were mainly in Dacaoping and Songluo village.

The mean FVC of Shennongjia Forest District between 1998 and 2013 was 66.8%, which was lower than 88.7%, the FVC of Hubei province, reported by Hubei Provincial Bureau of Statistics and Huang Zhe (Liu et al., 2018). One of the reasons is that the FVC estimated by remote sensing did not include the area of forest patches and gaps in forest, which was included in the provincial statistics. The second reason is that the proportion of deciduous forest inside National Nature Reserve is 61.0%, which is significantly higher than the 42.8% outside the reserve. Because of the defoliation, the FVC is low in non-growing season, so that the annual average FVC is underestimated (Liu et al., 2018). The higher FVC areas are mainly located in Yinyuhe, Jiuchonghe, Lichahe, Wenshuiliu, and Songluohe. Shennongjia Forest District is the only existing primary forest distribution area in the middle latitude of the world. The primary forest is mainly distributed in the high FVC area of Yinyuhe and Jinhouling. There is also a small amount of primary forest in Wenshuihe and Songluohe. The lower FVC areas are mainly around Dajiuhu, Songbai, Laojunshan, and the tourist roads inside the National Nature Reserve. There are large areas of crop lands, wetlands, and some bare rocks in Dajiuhu, causing the lower FVC. In Laojunshan and the northwestern of tourist roads inside the National Nature Reserve, the vegetation types are mainly alpine meadows and brush due to the high altitude, causing the lower FVC. Because of the provincial road 307 leading to Songbai, and the surrounding residential area, FVC is low within 3 km of both sides of road 307, which indicates that a human activity area such as road and residential area have influence on FVC.

The altitude of National Nature Reserve is between 400 m and 3100 m, covering the highest FVC area of altitude 1500 m to 2000 m. According to the results of Zhao et al. (2005), the high FVC is significantly correlated with the high number of arbor species and total species, indicating that the boundary of National Nature Reserve used to define the range of the World Natural Heritage Site is integral for the protection of biodiversity. It also shows that remote sensing is a good tool to estimate biodiversity of terrestrial ecosystem. The change of FVC shows that there were both significantly increase and decrease of FVC between 1998 and 2013 in Shennongjia Forest District. However, the total change rate was 1.45%, showing a growth trend of vegetation. By using Landsat TM data to classify the land cover type, Jiang (2015) found that the net increase of forest area of Shennongjia was 207.49 km^2. Huang & Xia (2013) also reported an increase trend of vegetation index of Shennongjia Forest District between 2003 and 2012 by using Modis EVI data.

The increase of FVC inside National Nature Reserve is 2.26%, twice of that outside the reserve, indicating the good effect of Natural Forest Protection Project in the reserve since 2000 (Liu et al., 2018). The possible reason of the FVC increase on the top of Laojunshan may be the plantation of *Larix kaempferi* forest there in 2000. The FVC increase in Qingshu village and Dongxi village located in the north of the National Nature Reserve is due to the removal of villagers in 2001, after which the vegetation was recovering. The reason of the FVC reduction around Dacaoping and Songluo village may be the construction of Shennongjia Airport and the ski resort. The influence of tourist roads and the provincial road 307 on FVC is complex, with both increase trend and decrease trend exist. It shows that roads can affect FVC, and are hoped to be effectively controlled in the future.

7.2.2.3 Factors Driving Forest Coverage Increase

We analyzed the relationships between the FVC change of Shennongjia Forest District and the nearest distance to the residential area and roads, as well as the relationships between FVC change and mean temperature, precipitation and land use type. The results show that there is a significantly positive relationship between FVC change rate and altitude, temperature, precipitation, and the nearest distance to residential area (altitude: $r = 0.245$, $p = 0.000$; temperature: $r = 0.101$, $p = 0.000$; precipitation: $r = 0.193$, $p = 0.000$; the nearest distance to residential area: $r = 0.031$, $p = 0.049$), while a negative relationship between FVC change rate and the nearest distance to roads ($r = -0.130$, $p = 0.000$) (Table 7.1) (Liu et al., 2018). Stepwise regression analyses show that the most important factors of affecting FVC change are altitude, temperature, precipitation, and the nearest distance to the roads. The nearest distance to the residential area also has impact on the FVC change rate. During 1998 and 2013, the significantly positive relationship between FVC change rate and altitude, temperature, precipitation, and the nearest distance to residential area, and the significantly negative relationship between FVC change rate and the nearest distance to the roads indicated that human activities have increased or decreased the vegetation in Shennongjia Forest District, specifically in the decrease around residential area and the increase near roads. There is a significantly decrease trend of FVC on the both sides of the provincial road 307, while an increase trend on the both sides of the provincial road 209. The reason is that the area around provincial road 209 has completed a 15.8 km^2 greening project in 2008 with the construction of Western Hubei Ecological and Cultural Tourism Circle. This result is

Table 7.1 The relations between the changing rate of FVC and influence factors

	FVC change rate	Nearest distance to the residential	Nearest distance to the road	Altitude	Temperature	Precipitation
FVC change rate	1					
Nearest distance to the residential	0.031*	1				
Nearest distance to the road	−0.130**	0.432**	1			
Altitude	0.245**	0.515**	0.115**	1		
Temperature	0.101**	−0.053**	−0.108**	−0.175**	1	
Precipitation	0.193**	0.159**	0.094**	0.387**	−0.041*	1

* = 0.05; ** = 0.01

the same as the research of Pauleit (2005) finding that urbanization manifests the increase of artificial green space and the decrease of natural vegetation.

7.2.3 Factors Driving Snub-Nosed Monkey Habitat Increase

The establishment of the Shennongjia Nature Reserve had promoted snub-nosed monkey's suitable habitat area increase by 17.70%. The most suitable habitat area increased by 8.88% (28.78 km^2) after the establishment of the reserve (1990), compared with that before the establishment (1980) (Wang et al., 2017). After the implementation of NFCP and SLCP (2010), snub-nosed monkey's suitable habitat increased by 14.29% (50.43 km^2) compared with that before the implementation (1990) (Fig. 7.5). It was also found that fragmentation of snub-nosed monkey's habitat significantly mitigated after the establishment of the reserve (Fig. 7.5). The metapopulation capacity of the habitat increased by 515.17% in 2015 since the reserve establishment (1980), and the metapopulation capacity of suitable habitat increased by 183.77% compared with that in 1980. The metapopulation capacity of the most suitable habitat reduced by 36.14% in 1990 compared with that in 1980, and the metapopulation capacity increased by 148.12% (Wang et al., 2017). The metapopulation capacity increased by 367.20% in 2010 after the implementation of NFCP and SLCP as compared with that in 1990 before the implementation.

The effectiveness in conserving different species is different in protected areas. It was found that the population and distribution range of the *Nimbaphrynoides occidentalis* showed an annually declining trend in the Nimba Mountains Nature Reserve as the World Heritage Site (Sandberger-Loua et al., 2016). Meanwhile, the Qionglai Mountains Nature Reserve only conserved 36% of the giant panda habitat

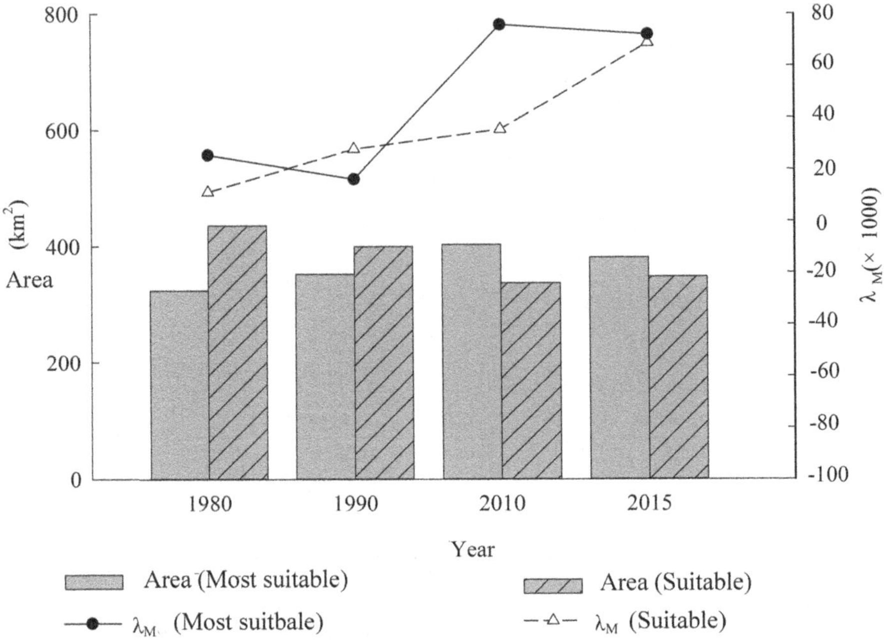

Fig. 7.5 The suitable habitats for Sichuan snub-nosed monkey since establishment

(Xu et al., 2006), while 45% of giant panda core habitat was outside the network of the reserve (Shen et al., 2008). As the endemic primate in China, Sichuan snub-nosed monkey is an endangered (EN) species of the IUCN Red List, the animal in the Appendix I of CITES and the primary national protected wild animal. At present, there are three subspecies, which are *Rhinopithecus roxellana roxellana* distributed in Gansu and Sichuan, *R. roxellana qinlingensis* in Shaanxi and *Rhinopithecus roxellana hubeiensis* in Shennongjia Nature Reserve and its surrounding areas (Li, 2007). Among them, Shennongjia Nature Reserve is only habitat for *Rhinopithecus roxellana hubeiensis*. With more than 30 years of strict protection and management, the population of the snub-nosed monkeys increased significantly, reaching up to 1282 in 2006 (Tie, 2013). Moreover, it was found that snub-nosed monkey's suitable habitat increased by 17.70% during more than 30 years after the establishment of the reserve.

It is of paramount importance to determine the assessment factors of habitat suitability (Shen et al., 2008). The indicator system was constructed for assessing habitat suitability of Sichuan snub-nosed monkey in Shennongjia, including vegetation, elevation, slope, aspect, and distance from roads and villages, among which the data of the vegetation were obtained based on remote sensing image interpretation with accuracy of 84%. The uncertainty of remote sensing image interpretation has a certain impact on the classification of habitat suitability. Shen et al. (2008) analyzed the uncertainty of the habitat suitability assessment of giant panda in Minshan Mountain, suggesting that the habitat suitability assessment was not sensitive to the uncertainty of remote sensing image interpretation when the accuracy of land cover type was over 80%, which meant that the errors in classifying the land cover type did not have a significant impact on the habitat suitability assessment. Habitat fragmentation has become one of the main factors of biodiversity loss (McGuire et al., 2016). Currently, landscape indicators have been analyzed in most studies of habitat fragmentation, such as the area of habitat patches, the number of patches, isolation distance, and shape index (Shen et al., 2008). However, these general indicators are limited to some extent, for they only focus on the spatial characteristics of habitat patches and fail to consider the biological characteristics of the species, and the results based on the different indicators are often inconsistent (Hanski & Ovaskainen, 2000; Schnell et al., 2013). Metapopulation capacity integrates the spatial characteristics of the habitat and the biological characteristics of the species, which well reflects the survival ability of species in the patches (Schnell et al., 2013). In this study, the habitat fragmentation degree of Sichuan snub-nosed monkey was analyzed through metapopulation capacity, based on the integration of the spatial characteristics and biological characteristics in the Nature Reserve, including the size of habitat patches and isolation distance, as well as the dispersal capacity of the monkey. Results suggested that the metapopulation capacity of this reserve increased by 515.17% in the past 30 years and the fragmentation degree was significantly reduced, which showed that Shennongjia Nature Reserve had significant effect in conserving the flagship species and its habitat.

During the 10 years (from 2000 to 2010) after the implementation of NFCP and SLCP, the ecosystem service function in China has generally improved, with the net increase in forest area of 120,000 km^2 and the increase in the food production of 38.5%, the carbon sink of 23.4%, soil retention capacity of 12.9%, and flood disaster reduction capability of 12.7% (Ouyang et al., 2016). Commercial logging was once a major threat to snub-nosed monkey's habitat. The results suggested that the forest area, forest coverage, and the suitable habitat for snub-nosed monkey showed an increasing trend after the project of NFCP and SLCP, which further improved the conservation effectiveness of Shennongjia Nature Reserve.

The establishment of Shennongjia Nature Reserve has mitigated the forest loss and habitat degradation, increased the forest coverage and suitable habitat of snub-nosed monkey, and significantly decreased the degree of habitat fragmentation, indicating that the reserve achieved good results in conserving the forest and the habitat of Sichuan snub-nosed monkey. Moreover, the NFCP and SLCP further enhanced the conservation effectiveness of Shennongjia. However, the Shennongjia Nature Reserve is still threatened by firewood collection and development of tourism and forestry (Xiang et al., 2011). The conservation effectiveness of protected areas is the result of the combined effect of various factors (Gaston et al., 2008). The interaction of social and economic factors inside and outside the protected areas should be taken into consideration, and the future trends of the factors and their possible influence on the conserved objects and conservation effectiveness should be predicted in the future assessment of conservation effectiveness of Shennongjia Nature Reserve. Meanwhile, it is necessary to strengthen cooperation between research institutions, local governments, and communities to jointly improve the conservation effectiveness of the reserve.

7.2.4 *Environmental Pressures and Natural Disasters*

Since Hubei Shennongjia has been designated as a provincial-level scenic spot, a National Nature Reserve, and a Global Geopark, some pressures from ecosystem degradation have been effectively mitigated. The increasing tourists have cast some ecological pressures on Hubei Shennongjia. The activities such as tourist visitation, vehicles pass through, and the construction of reception facilities would bring some deterioration to the quality of air, water, and induce noise pollution. The authority has implemented corresponding measures to mitigate the impact.

The main threat from natural disasters to Hubei Shennongjia includes geological disaster, frozen disaster, forest fire, forest toxic organisms, as well as ecological disasters due to global warming. The administrative authorities are taking counter-measures to protect Hubei Shennongjia against the disturbance by natural disasters. The authorities have established a warning system to monitor and prevent the disasters. The authorities should be established for organization and responsibility systems to monitor and prevent geological disasters. There was lots of training for protection or fire prevention. There are warning boards informing people in the area not to smoke and not to use flammable materials. The forest pests and diseases monitoring and associated research have been carried out to forecast and monitor forestry pests and diseases.

7.2.5 *Pressure from Tourism Development*

A small amount of general tourism, as well as some adventure tourism, hiking, and scientific investigation, exists in Hubei Shennongjia. The eco-tourism area in Hubei Shennongjia is large, and the distance between scenic areas is long, therefore environmental capacity estimation should be based on scenic area. Because scenic spots in each scenic area are scattered, the sightseeing of most tourists is mainly along the road in scenic area, thus tour route method is applied in calculating the environmental capacity in each scenic area. Currently, there are four scenic areas in Hubei Shennongjia, namely Shennong Peak and Guanmen Mountain. The daily environmental capacity of Shennong Peak scenic area is 7000 visitors, and the daily environmental capacity of Guanmen Mountain scenic area is 3200. Totally, the daily environment capacity of the Shennongjia is 10,200. The tourist season in Hubei Shennongjia is from May to October every year. Tourism peak is in weekend days and National Day holiday. Hereafter the annual tourism day is calculated from the tourist season as 180 days. Thus, annual environmental capacity is 1.83 million of people (Yu et al., 2018b).

Table 7.2 Population located within Hubei Shennongjia and its buffer zone

	Area (ha)	Population
Area of property	73,318	6999
Buffer zone	41,536	7388
Total	114,854	14,387

Following the construction and improvement of eco-tourism areas, the exploration of new scenic attractions, the advertisement and development of tourism market, this rapid development stage of eco-tourism in the property persists for some times. Therefore, the number of tourists should be below the limitation by the annual environmental capacity during the planning period. However, in the tourist season, tourists may be concentrated in scenic areas, resulting in the saturation of the environmental capacity, in that case, the number of tourists shall be controlled strictly to minimize the adverse influence from human activities on the natural environment. The number of tourist arrivals had been limited through a system. Tourists are only allowed to enter the hotspots one group at a time. A rotation system had been adopted, where tourists can only visit certain spots at any one time.

7.2.6 *Pressure from Community Development*

The total population of Shennongjia in 2011 was 79,000. These people are located in the region's six towns. The area's main industries are eco-tourism and the forest tea industry. There are 6999 local people in Hubei Shennongjia, meaning it has a population density of 10.4 people per square kilometer. This is less than the 25 people per square kilometer taken as the measure of a sparsely populated area (Table 7.2). There are several towns around Hubei Shennongjia, including Muyu town, Xiagu town, and Jiuhu town. In order to protect the ecological environment and rare flora and fauna of Hubei Shennongjia, the government has accelerated the economic transformation and upgrading of agriculture in the area. In doing so, it aims to reduce the damage caused to the resources and environment by rural economic development. Measures have been taken to reduce the burden on farmers and increase farmers' income including the development of characteristic agriculture and forestry, the development of eco-tourism and related services, and the implementation of ecological compensation schemes.

Chapter 8
Protection of Hubei Shennongjia

8.1 Protection of the Flagship Species and Key Biodiversity

Hubei Shennongjia at different levels of the administrative structure is needed to be protected.

8.1.1 Sichuan Snub-Nosed Monkey

Hubei Shennongjia is the current distribution area of the subspecies of Sichuan snub-nosed monkey. The latest survey shows that there are 1550 snub-nosed monkeys in eight groups. They mainly live in Golden Monkey Ridge, Da Longtan, Xiao Longtan, Da Qianjiaping, and Xiao Qianjiaping, with an altitude ranging from 1600 to 3000 m. Recent surveys show that the main distribution area in Hubei Shennongjia is Golden Monkey Ridge sub-area (about 800 individuals), and Qianjianping sub-area (about 400 individuals) (Quan & Xie, 2002), there are a few small groups or individuals in Baiyanpo and Banbiyan. In the main activity area of Sichuan snub-nosed monkey, the tourism road in Da Longtan can affect monkeys' movements and intra-specific communication. To facilitate the movement and genetic communication, two snub-nosed monkey crossings should be constructed in the planning (Fig. 8.1). The tourism road in Da Longtan laid through high mountains and deep valleys, constructed on the karst geomorphology. Therefore, it is difficult to build below-ground crossing.

Based on field surveys and the long-term monitoring data for snub-nosed monkey, we proposed the details of the crossings for Sichuan snub-nosed monkey as follows: (1) Wildlife overpass is suitable for the success passing of Sichuan snub-nosed monkey and best to be constructed in Dalongtan and Xiaolongtan considering their population distribution, and overpass with 80 m width can increase the probability of success passing for the Sichuan snub-nosed monkey (Zhou et al., 2017a);

Z. Xie, G. Shen, *The Outstanding Universal Value and Conservation of Hubei Shennongjia*, https://doi.org/10.1007/978-981-16-0684-7_8

Fig. 8.1 Schematic diagram of crossing for Sichuan snub-nosed monkey

(2) The substrate of overpass should be similar to the surrounding environment, and would be better if the soil meets 2.2–2.4 m depth which is essential to the growth of plants; (3) Plants that snub-nosed monkeys prefer (*Abies fargesii, Usnea longissima, Betula albosinensis, Carpinus cordata, Decaisnea insignis, Fagus engleriana*) should be cultivated; (4) Infrared monitoring system should be set up in the middle and the entrance of the overpass, and the monitoring data should be transferred automatically; (5) Other facilities like fence along the both sides of the entrance of overpass with 2.4 m height, vine-covered protective wall, and warning signs ("attention wildlife," "wildlife overpass") should be erected to ensure the effectiveness of wildlife crossing (Yu et al., 2018a). We suggest that the benefits acquired from wildlife crossings been maximized with continuous monitoring, well preserving, and in accordance with the improvement of public conservation awareness of both eco-tourists and tour operators driving within the reserve. The efficient use of these crossings may make a great contribution to improve the balance between development and biodiversity conservation in the nature reserve (Yu et al., 2015).

8.1.2 Rare and Endangered Plants

Rare and endangered species in Hubei Shennongjia are mainly distributed in damp valleys and several special areas characterized by little human disturbances. The main protection measures for rare and endangered plants include setting up in situ conservation sites in the six key areas, raise the seed setting rate of rare/endangered plants through appropriate methods, to promote their population regeneration and rejuvenation. Establish nurseries to ex situ conserve rare and endangered plants through appropriate introduction and propagation methods. At the same time, strengthens studies, experiments, and observations on the biological and ecological characteristics of rare and endangered plants, to provide the theoretical basis and technical support for conservation.

The six key distribution areas including Jiuchong river basin, Yangquan river basin, Yinyu river basin, Changping, Majia river, and Shashuwan area. Jiuchong, the key habitat for *Taxus wallichiana* var. *mairei, Davidia involucrata, Manglietia patungensis, Michelia martini,* orchid plants such as *Dendrobium* sp., kiwi plants such as *Actinidia rubricaulis* var. *coriacea.* Yangquan river basin, the key habitat for *Cercidiphyllum japonicum, Torreya fargesii, Taxus wallichiana* var. *chinensis*, orchid plants such as *Cypripedium japonicum* and *Pleione bulbocodioides*, kiwi plants such as *Actinidia chinensis* and *Actinidia kolomikt*a. Yinyu river basin, the key habitat for *Davidia involucrata, Cercidiphyllum japonicum, Taxus wallichiana* var. *chinensis, Paris polyphylla*, kiwi plants such as *Actinidia callosa* var. *henryi* and *Actinidia tetramera*, orchid plants such as *Goodyera biflora*. Changping river basin, the key habitat for *Taxus wallichiana* var. *chinensis, Torreya fargesii, Cercidiphyllum japonicum, Huperzia serrata,* orchid plants such as *Cypripedium franchetii*, kiwi plants such as *Actinidia chinensis* var. *deliciosa*. Majia river basin, the key habitat for *Fagus hayatae, Davidia involucrata, Taxus wallichiana* var. *chinensis*, orchid plants such as *Bletilla ochracea*, kiwi plants such as *Actinidia tetramera* var. *badongensis*. Shashuwan area, the key habitat for *Davidia involucrata, Taxus wallichiana* var. *chinensis, Clethra fargesii*, kiwi plants such as *Actinidia chinensis*, orchid plants such as *Calanthe discolor*.

8.1.3 Primeval Forest

The primeval forest is defined as the mature or over-mature forest, which is rarely disturbed by human activities. Primeval forests in Hubei Shennongjia have conserved various endangered species of the North Subtropical Zone, ensured the integrity of the eco-environment of Hubei Shennongjia, and guaranteed the best protection of local biodiversity. Further, the primeval forests maintain the natural ecological system evolution and ensure the protection of natural landscapes, biological habitats in Hubei Shennongjia, effectively facilitating the ongoing biological and ecological processes of communities and ecosystems of plants and animals. Hubei Shennongjia is sparsely populated, and relative intact primeval forests of an area of 17,365 ha are well preserved. The main measures to protect primeval forests are: long-term monitoring through remote sensing, evaluating the scientific achievements of protection and management, as well as the succession of forest ecosystem; felling and logging are strictly prohibited in the area; carry out the defense activities regularly to prevent poaching and illegal logging; strengthen the mechanisms for prevention of forest fire and pest, improve the rapid response system of early warning, and prevent the destruction of the primeval forest by natural disasters. At the same time, the destructive behavior was punished legally.

The protection objects of Shennongjia, the World Natural Heritage Site, are to be strictly protected by national and local laws and regulations. However, large-scale forest harvesting was carried out in the Shennongjia Nature Reserve in the 1970s, which severely damaged the habitat of snub-nosed monkeys. With the rapid development of Shennongjia tourism and the annual increase in the number of tourists in recent years (Xiang et al., 2011), the reserve still faces the threat of firewood collection and invasion of tourism and forestry, which leads to the prominent contradiction between protection and development.

8.1.4 Mountain Altitudinal Vegetation

Within Hubei Shennongjia, the topography with a vertical elevation of between 420 and 3100 m was saved from the damaging effects of the Quaternary glaciers and the blast of cold wave from Mongolia-Siberia. This area has also benefitted from the southeast monsoon and southwest monsoon. Hubei Shennongjia is rich in flora due to the fact that it is a convergence zone between subtropical flora and warm temperate flora. There is an elevational vegetation belts within Hubei Shennongjia. Hubei Shennongjia plays a key role in the shifting, exchanging, mixing, and evolution of the flora from Sino-Himalaya and Sino-Japanese. It is the high concentration of all of the subtropical, warm temperate, temperate, and cold temperate ecosystems in such a small area that makes Hubei Shennongjia an outstanding example for research into the ecological processes of mountain ecosystems along the elevation gradients under global climate change. There are two areas in Hubei Shennongjia which have relative intact and well-preserved altitudinal natural zones, with a total area of 20,760 ha. One area is located on the south slope of mountains in Jiuchonghe, another is located on the mountains in Yinyuhe. Main measures to protect altitudinal natural zones are: set up long-term monitoring plots at the two sites mentioned above, and regularly monitor the vertical variation of mountain ecological processes by global climate change. Felling and logging are strictly prohibited in the two areas. The destructive behavior was punished legally.

8.1.5 Mixed Broad-Leaved Evergreen and Deciduous Forest

The subtropical mixed broad-leaved evergreen and deciduous forest in Hubei Shennongjia is the best one in the Northern Hemisphere. It presents a typical case for elevation vegetation in the Oriental Deciduous Forest biogeographical province. There are three large areas in Hubei Shennongjia which well preserve the mixed broad-leaved evergreen and deciduous forest, with a total area of 4250 ha. Among them, the Jiuchonghe site is composed of *Lindera* species of Lauraceae (evergreen component) and *Fagus* species of Fagaceae (deciduous component). While the Changpinghe site is composed of *Eurya* species of Theaceae (evergreen component) and *Carpinus* species of Betulaceae (deciduous component). The Yangquanhe site is composed of *Litsea* and *Lindera* species of Lauraceae (evergreen component), as well as *Quercus* and *Fagus* species of Fagaceae (deciduous component). The main measures to protect mixed broad-leaved evergreen and deciduous forest are: set up long-term monitoring plots at the three sites mentioned above, and regularly monitor the ecological processes of mixed broad-leaved evergreen and deciduous forest. Felling and logging are strictly prohibited in the three areas. The destructive behavior was punished legally.

8.2 Environmental Protection

8.2.1 Water Environment

Hubei Shennongjia is located in the Middle Route in the South-to-North Water Diversion Project. In this area, the protection of the water environment is considered to be of high importance. In Hubei Shennongjia, there are many rivers and streams with clean, crystal clear water.

Water resources are an important part of the landscape resources within the scenic region. They are also a source of negative oxygen ions and drinking water. All of the domestic sewage within Hubei Shennongjia must be discharged after complying with prescribed treatments. To prevent water pollution from soil erosion in the scenic region, mining and quarrying are prohibited in Hubei Shennongjia. At the same time, planting and restoring vegetation is requested after the construction of paved footpath. To avoid water pollution, materials that are susceptible to leaching should not be stacked next to rivers. People are prohibited from throwing debris such as garbage, peel, paper, and so on into the rivers.

8.2.2 Atmospheric Environment

Hubei Shennongjia has 96% forest coverage and is far away from industrial regions. The air quality is fresh throughout the whole year. The percentage of days where the ambient air quality is good each year is 100%. The number of days with good readings on the ambient air pollution index is more than 85%. Air quality is up to Class I of the national *Ambient Air Quality Standard (GB 3095-1996)*. There are 166,000 negative oxygen ions per square centimeter in the air of the scenic region where Hubei Shennongjia is located, its forests therefore offering a natural oxygen bar.

The air quality in Hubei Shennongjia can be classified as first-class. In Hubei Shennongjia and its buffer zone, the ambient air quality should meet the first-class standard on 99% or more of the days in each year. In peripheral area of Hubei Shennongjia and its buffer zone, ambient air quality meets the standard of 95% or more. Measures including the use of fuel to provide heat in people's homes should be replaced by the use of liquefied gas, electrical energy, solar energy, and biogas. The quality of gas appliances should reach high standards. Environmentally friendly vehicles should be used as traffic tools for tourists in Hubei Shennongjia. The administration of Hubei Shennongjia should endeavor to monitor air quality carefully, to do dynamic analysis regularly, and when air pollution is detected they should take measures to rectify problems in a timely fashion.

8.2.3 Sound Environment

There are no loud retailers in Hubei Shennongjia. Environmentally friendly vehicles such as electro-mobiles are used, which produce little noise According to an assessment by the environmental protection department, the sound environmental qualities reach the Class 1 standard of the national requirement. Currently, the Shennongjia airport (11 km away from Hubei Shennongjia) has begun operation. The monitoring of the noise shows that the disturbance range of the noise with more than 60 dB effects from the Shennongjia airport was 4.4 km^2 (Table 8.1). According to the "Environmental impact report of the Shennongjia civil airport project," the aircraft with south-eastern route fly over Hubei Shennongjia, with a height of over 800 m above the ground, and the noise value is lower than 50 dB, causing little impact to animal activity in the protected area.

The noise control in Hubei Shennongjia meets the Class 1 standards, with the noise less than 55 dB (A) in the day and 45 dB (A) at night. The noise created by any service facilities in Hubei Shennongjia should not be above the national standards. Sound control measures including the management of the area seek to eliminate and reduce noises, by controlling the sound sources. As such, loudspeakers are prohibited in the scenic region. The manager should use sound insulated facilities and construct a sound insulating forest belt to prevent noise. The environmentally friendly shuttle buses and the protection vehicles are allowed to enter Hubei Shennongjia's display district. The administration irregularly monitors the footpaths and tourists in the exhibition district. Under the premise of safety, airlines should ensure that they do not fly at low altitude when flying over Hubei Shennongjia for controlling the noise level.

Table 8.1 The range of the noise from the airplane in Shennongjia airport (Unit: Km^2)

>60 dB	>65 dB	>70 dB	>75 dB	>80 dB
4.4	1.8	0.9	0.6	0.5

8.2.4 *Soil Environment*

Protection measures are transfering solid waste to outside of the scenic region rather than be discarded randomly. It also should not be buried in the soil directly. Measures should be taken to prohibit the cutting into mountains for quarrying and for obtaining sand and soil. The administration should prohibit the construction of large buildings and architecture. The amount of chemical fertilizer and pesticide used on farmland around Hubei Shennongjia should be controlled. Measures should be taken to prevent agricultural non-point source pollution and soil pollution.

8.2.5 *Environment Sanitation*

Due to Hubei Shennongjia far from the towns, that sewage is discharged as standard levels after ecological treatment based on land-third-level-filter. Classified garbage bins were placed in Hubei Shennongjia. Garbage bins (or litter bins) should be made of environmentally friendly materials and should allow trash to be separated into recoverable/recyclable and unrecoverable/unrecyclable materials, with labels. Facilities for garbage treatment should be sufficient and intact. Garbage should be put into bags each day. The garbage in bags should be transferred to refuse disposal plants in a timely way, and be treated by burning and burying. The excrement from restrooms should be decontaminated and disposed of. There should be sanitation staff working within Hubei Shennongjia. They should clean toilets and keep areas clear from waste. Sanitation vehicles should be used to remove garbage based on a daily garbage production. All sanitation equipment should be designed to be attractive and clean in accordance with the surrounding environment.

8.3 Connection of the Two Isolated Components

Hubei Shennongjia, as a World Natural Heritage Site with exceptional rich biodiversity and significantly unique ecological and biological processes, offers an outstanding example of the biodiversity, ecosystem, and ongoing ecological processes in the Northern Hemisphere. Shennongjia World Natural Heritage Site composes of two components, which at present are ecologically separated by quite a large distance including a fenced National Road 209.

8.3.1 *Feasibility of Enhancing the Connectivity*

Due to human activities, Hubei Shennongjia is composed of two disconnected components, with the nearest interval of 3 km and the farthest 13.4 km. In addition, the National Road 209 passes through the buffer zone between the two components. This buffer zone covers 232 km^2, ranging from 740 m to 2900 m. With the implementation of several protection projects (e.g., the Natural Forest Protection Program, the Grain for Green Project) over past years, the buffer zone between the two components has been protected and rehabilitated. Currently, vegetation covers 95% of the buffer zone. The well-preserved buffer zone provides the feasibility for enhancing the connectivity between the two components.

8.3.1.1 Requirements for Establishing the Wildlife Corridor

Ecological corridor has been proposed as essential for restoring connectivity in the fragmented landscape. The width of the ecological corridor, which is generally decided based on the demands of the largest target species, determines its efficiency (Chetkiewicz et al., 2006). When the corridor width is larger than the minimum movement distance of the largest target species, the corridor can facilitate the movement of almost all species and relating ecological functions (Pérez-Hernández et al., 2015). Scientific researches indicate that corridors can meet the movement requirements of large and medium sized mammals when the width is greater than 1.2 km (Silveira et al., 2014). To allow for the greatest degree of wildlife movement, the minimum width of activity of the largest fauna species is used to set the minimum width of the corridor, the diameter of the circle of activity (Zhu et al., 2005).

The largest animal in Hubei Shennongjia is the Asian black bear (*Ursus thibetanus*), and its core home range size is 1.4 km^2 (Trent, 2010). This size is regarded as the circle of activity, and its diameter thus is 1332 m. With low human disturbance in the northern and southern areas of the buffer zone between the two components, the vegetation covers over 97% of the area and the habitat quality is very high. The minimum width of continuous distribution of natural vegetation is 2.5 km and 1.4 km in the northern and southern areas, respectively, which meets the minimum width requirement for the wildlife corridor. Notably, the southern area of the buffer zone between the two components includes an entire watershed area, which should be included in the proposed wildlife corridor to sustain the movement of aquatic animals and amphibian activating in mid or low altitude regions (Zhu et al., 2005). Although the distribution of natural vegetation, with sparse, small patches of farmlands (0.12 km^2, accounting for 0.34% of this area) and settlements (31 households), is not continuous within this watershed, the habitat quality is high, and the connectivity can be improved through emigration and habitat restoration, thus meeting the requirements for establishing the wildlife corridor.

8.3.1.2 Requirements for Building Stepping Stone

Stepping stones have been proposed as an auxiliary means of promoting the connectivity of fragmented landscapes. The area and interval of the stepping stones are two key factors in determining its functional efficiency. The area of the stepping stone must be larger than or equal to the minimum area required by the largest target species, and so the minimum area of the stepping stone is determined by the core home range size of the largest target species. The interval of the stepping stones (the distance between adjacent stepping stones, or between it and protected area) is determined by the minimum daily movement distance of habitat specialist species (Hill et al., 1993).

The largest animal within Hubei Shennongjia is the Asian black bear (*Ursus thibetanus*), its core home range size is 1.4 km^2 (Trent, 2010), which is regarded as the minimum area of the stepping stone. The elevation of the buffer zone between the two components ranges from 740 m to 2900 m. Specialists distributing within this elevation zone are mostly large and medium sized mammals. Among them, the yellow throated marten (*Martes flavigula*) is the smallest species, and its daily movement distance is 2051 m (Grassman et al., 2005), which is considered as the largest interval distance between stepping stones. Aside from sparse, small patches of farmlands and settlements, the vegetation coverage in the middle area of the buffer zone between the two components is 97.23%, dominated by natural forest (Table 8.2). Most of these patches without settlements and farmland are over 1.5 km^2, and the intervals between adjacent patches are less than 1 km, therefore meeting the requirement of minimum area and maximum interval for establishing stepping stones.

8.3.1.3 Wildlife Crossing Construction and Road-Fence Dismantlement

Wildlife crossings, including overpasses and underpasses, have been proven to be an effective way to alleviate the isolating impact of roads. Overpasses, mainly constructed in U-shaped sites, are suitable for terrestrial mammals and a few amphibians to cross roads in safety. Underpasses, constructed in wetlands and streams, are suitable for aquatic animals, a few amphibians and terrestrial mammals (Beckmann et al., 2010). The length and width of the National Road 209 where it crosses the buffer zone are 32 km and 10 m, respectively. This fenced road hinders potentially animal movement. High quality habitats on both sides of most sections of the road can facilitate animal crossing after the fence dismantlement. The junctions of the proposed corridors and the National Road 209 are U-shaped topography, which meet the requirements for building wildlife overpasses to enhance wildlife movement. Seven wildlife underpass crossings have already been built at the junction of the stream and the National Road 209, which facilitate movement of aquatic animals, amphibians, and terrestrial animals.

Table 8.2 Vegetation cover in the buffer zone between the two components of Shennongjia World Natural Heritage Site

Forest type	Buffer zone between the two components	North corridor	South corridor middle area of the buffer zone between the two components	Stepping stone				
				T1		T2	T3	T4
Broadleaf evergreen forest	0.02	0.00	0.00	0.03	0.00	0.00	0.00	0.00
Mixed broadleaf evergreen and deciduous forest	50.19	56.86	57.35	43.50	62.32	89.92	3.16	9.07
Broadleaf deciduous forest	35.91	15.44	36.98	48.29	37.68	9.88	92.15	90.93
Mixed needleleaf and broadleaf forest	4.62	12.60	0.00	1.28	0.00	0.00	0.00	0.00
Subalpine needleleaf forest	0.00	0.00	0.00	0.00	0.00	0.00	0.00	0.00
Needleleaf forest	5.65	14.13	2.70	1.40	0.00	0.20	0.00	0.00
Scrub	0.85	0.09	0.12	1.58	0.00	0.00	0.00	0.00
Total	97.23	99.11	97.14	96.09	100.00	100.00	95.32	100.00

8.3.2 Implementation Scheme

The State Party China had established landscape mosaics with wildlife corridors, stepping stones, and crossings to enhance the connectivity for ecosystems and species between the two components. The State Party China had built two wildlife corridors at the south and north area of the buffer zone between the two components, and four stepping stones between the south and north corridor. Over 30% of the fence along the National Road 209 across the buffer zone between the two components and its branch within the components had been dismantled. Two wildlife overpass crossings had been constructed in each corridor. This infrastructure, as well as seven established wildlife underpass crossings, enhances the connectivity of the two components.

8.3.2.1 Establishing Wildlife Corridor and Stepping Stone

The proposed corridor at the northern end of the buffer zone between the two components is 12 km in length, and 4.6 km and 2.5 km at maximum and minimum width, respectively. It covers 51.37 km^2, with an altitude range of 1700–2900 m and natural vegetation cover of 98%. The corridor effectively facilitated the exchange of wildlife and associated ecological functions between the two components in the mid or high altitude zones. The proposed corridor at the southern end of the buffer zone between the two components is 8.5 km in length, and 5.3 km and 2.7 km at maximum and minimum width, respectively. It covers 35.86 km^2, with an altitude range of 900–2700 m and natural vegetation cover of 97% (Table 8.2). The corridor effectively facilitated the exchange of wildlife and associated ecological functions between the two components in mid or low altitude zone. Four stepping stones had been built between the two corridors. Their areas are 3.45 km^2, 1.98 km^2, 1.85 km^2, and 1.53 km^2, respectively. The intervals of two adjacent stepping stones are less than 1 km.

8.3.2.2 Wildlife Crossing Construction

Two wildlife overpass crossings had been constructed within each corridor. A total of 11 crossings, including the seven already established underpass crossings had effectively alleviated the barrier effect of National Road 209. To alleviate the barrier effect of the branch road on the Sichuan snub-nosed monkey, two wildlife overpass crossings have been previously proposed at the branch of the National Road 209 within the Shennongding-Badong component. The essential width of overpass crossing is 60–80 m. The substrate for the wildlife crossing should be similar to the surrounding environment, and the solid should ideally be 2.2–2.4 m in depth, which is necessary for plant growth. The crossing had been planted with forest dominated by native species with a canopy cover of 70–80%. Meanwhile, vine-covered protective wall and warning sings (e.g., "attention wildlife," "wildlife crossing") had been erected to promote the effectiveness of the wildlife crossing. The width of the seven already established underpass crossings is 25–60 m, and several warning signs for wildlife conservation have been set in each established underpass crossing.

8.3.2.3 Road-Fence Dismantlement

Fence had been dismantled following four principles: (1) dismantling all fences in the corridor zones and high habitat quality zones; (2) partly dismantling fence in radiation zone of the stepping stones; (3) maintaining fences in high-intensity human disturbance zones; and (4) maintaining fences in steep areas with cliffs. Six sections (14.5 km) of fence along the National Road 209 had been dismantled, accounting for 45% of its total length. In addition, 11 sections (13.2 km) of fence along the branch of the National Road 209 within the Shennongding-Badong component are proposed to be dismantled, covering 30% of the total road. Considering the frequent movement and activity of mammals, amphibians, and reptiles during the night and the breeding season, the maximum speed of motor vehicles was been limited under 30 km/h in the sections that fences had been dismantled.

8.4 Protection Zoning

The decision about the type of protection given to these elements is based on each element's current situation, their distribution within the area. There are two kinds of districts in Hubei Shennongjia: *prohibited district* and *display districts*. Buffer zones have also been established in the area surrounding Hubei Shennongjia.

8.4.1 Prohibited and Limited District

The prohibited district is strictly protected for the heritage's prominent diversity and ecosystem, and most of human activities are prohibited other than Popular science education and ecological tourism. Some of the protection priorities in the area are to prohibit the construction of any environmentally incompatible buildings in the area. Tour facilities are limited to simple service points near walkways, signboards, sanitary facilities, awareness points, recreation facilities, maintenance stations (outposts). Large service facilities are strictly prohibited, including resorts, hotels, training centers, and sanatoriums. Except motor vehicle roads, other roads are not allowed to be built at will. All kinds of buildings and roads prohibited, without approval or not related to the protection of resources, should be relocated or within a limited time. Tourism and environmental monitoring in the region should be strengthened to ensure the sustainable use of resources. Except for environmentally friendly shuttle buses and vehicles used for management tasks, no car can enter the district.

8.4.2 *Exhibition District*

In this area, the infrastructure and service facilities required for tourism are allowed to be built (Fig. 6.10). This area is the main area providing short-term services to tourists. Projects not related to sightseeing are prohibited. Some of the area's protection focuses on motor vehicle roads and the necessary horses for sightseeing. Some facilities are permitted, including sidewalks, restrooms, awareness-raising facilities, and small service points. The construction project should be adapted to the environment and landscape of Hubei Shennongjia. In short, all activities about construction must be compatible with the natural landscape, and large-scale construction in the exhibition area is not allowed.

8.4.3 *Buffer Zone*

The function of the buffer is to protect the integrity of ecosystem and biodiversity for Hubei Shennongjia. In this area, the size and scope of motorized transport, resident population, and tea planting should be controlled in the appropriate range (Fig. 6.10). The buffer zone can implement some special ecological restoration activities. It is strongly discouraged to construct or destroy landscape, or pollute the environment.

8.4.3.1 Function and Scope of Buffer Zone

The buffer zone is a transitional zone, which separates the core area from external interference and aims to protect Shennongjia from human activities. In addition, the buffer zone can be protected and moderately developed to ensure that resources and the environment are protected (Fig. 6.10). The boundary of the buffer zone is determined mainly on the basis of: The area must be adjacent to the core area, it is part of the protected area to ensure the protection of the heritage value, and it must be large enough to protect the core area well.

8.4.3.2 Current State of Buffer Zone

The heritage's buffer covers 41,536 ha, inhabited by 7388 local people. Currently tourists coming to the area almost all stay in Muyu town in the buffer zone. The buffer zone of Hubei Shennongjia is also a protective area. As such, the Shennongjia National Park Authority undertakes protective measures to protect the heritage, and the center or province government provide funding for the Authority. Ninety-five percent of the buffer is covered by forests; the dominant forests are evergreen broad-leaved forests, mixed evergreen and deciduous broad-leaved forests, deciduous forests, mixed needle and broad-leaved forests, needle forests, and some shrubs. Due to the protection of Yanduhe Provincial Nature Reserve in Badong County for the boundary of Hubei Shennongjia, which is bounded on the south by Hubei Shennongjia, there is no buffer zone in the boundary adjacent to the Badong County. By the way, the buffer zone in northern part of the heritage is narrow because of the stiff landform and lower level human disturbance. So, all of the buffer zone played an active role in buffering Hubei Shennongjia against human encroachment. Although some part of the zone is narrow, it can protect Hubei Shennongjia from human disturbance due to the high mountains and low intensity of human activities.

8.4.3.3 Protection and Management

Different departments and groups work together to protect and manage the buffer zone. Forestry departments, public security departments, and villages and towns in the buffer zone are included in the defense system. The protection and management responsibilities of these groups are clarified. The functions of checkpoints in the buffer zone, including phytosanitary, pest control, forest fire control, and timber inspection, are detailed to how to do. In this way, the timely removal of unsafe factors from the buffer zone and the strengthening of cooperation between various departments and regions have created long-term effective mechanisms. It is encouraged for community to participate in buffer zone protection and management. Shennongjia National Park Authority raises public awareness of protection through television, newspapers, leaflets, and other media. At present, residents of the buffer zone have been helped to participate in the protection of public welfare forests, and local village residents have been invited to voluntarily form patrol teams to strengthen forest protection in the buffer zone.

8.4.3.4 Buffer Zone Management

The heritage management department is responsible for buffer zone planning and communicates and coordinates with stakeholders on behalf of the local government. The construction of the buffer zone must be carried out in accordance with the protection and management plan, and the construction of the buffer zone has been approved by the local government. Some production, operation and management facilities can be built in the buffer zone, but the scale should be strictly controlled. Land use activities in the buffer zone can not adversely affect the value of heritage. According to the protection and management plan, if there are differences between local government and stakeholder, the stakeholders have been consulted by the people's Government of Shennongjia Forestry District. The conservation of heritage sites has been strengthened through regular joint conservation actions. A joint protection system was established and the protection was extended to the entire buffer zone. Local residents were organized to protect heritage sites and subsidies, and incentives were implemented in accordance with the policy of protecting natural forests.

8.5 Managing Units in Hubei Shennongjia

Conservation stations are being established at different places across Hubei Shennongjia. These conservation stations (also conservation points) must control all of their designated area. The locations of these conservation stations (and points) in Hubei Shennongjia should be based on the properties different managing units, the heritage conservation status, and other facilities. The protection stations aim to prohibit the heritage against human disturbance to ensure the heritage integrity by monitoring, patrolling, and carrying out other protective measures. The conservation points are for fire control and the prevention of unlawful hunting. They are also places to conduct environmental monitoring and observation. The staff patrolling Hubei Shennongjia can get aid or food and drink here. Seven conservation stations and 15 conservation points had been established in Hubei Shennongjia (Fig. 6.10). These perform different functions, focusing on protecting different objects, as detailed below (Table 8.3).

8.6 Monitoring in Hubei Shennongjia

8.6.1 Monitored Objects

According to the heritage type and protection status of Hubei Shennongjia, a heritage management and monitoring system was established. The system comprehensively monitors the ecosystem integrity and biodiversity of heritage sites, and provides integrated monitoring of climate change, air quality, water quality, noise, solid waste pollution, and environmental health. At the same time, the monitoring system also monitors tourism activities and tourism service facilities and quality. Finally, the monitoring system monitors landslides, fires, and pests.

8.6.2 Improvements to the Monitoring Systems

8.6.2.1 Monitoring Center

Heritage management has established heritage science and digital information centers to improve scientific research and information processing capabilities. The Center is responsible for monitoring information on the protection and management of heritage sites and their buffer zones, regularly analyzing monitoring information,

Table 8.3 Protection stations (points)

The protection station	Point	Activities
Guanmenshan	Qianjiaping Zhangbaohe	The focus protection objects: Subalpine forest ecosystems, natural landscape, and key protected wild fauna and flora, such as *Rhinopithecus roxellana*, *Moschus berezovskii*, *Neofelis nebulosa*, *Panthera pardus*, *Aquila chrysaetos*, *Carabus* sp., *Davidia* sp., *Cercidiphyllum japonicum*, and *Emmenopterys henryi*, etc.
Yinyuhe	Bancang Yuergou Baibuti Yazikou	The focus protection objects: Subalpine forest ecosystems, natural landscape, and key protected wild fauna and flora, such as *Rhinopithecus roxellana*, *Andrias davidianus*, *Moschus berezovskii*, *Lutra lutra*, *Neofelis nebulosa*, *Panthera pardus*, *Ursus thibetanus*, *Aquila chrysaetos*, *Luehdorfia chinensis*, *Davidia* sp., *Cercidiphyllum japonicum*, *Emmenopterys henryi Tetracentron sinense*, etc.
Banqiao	Xiagu Bailinyan	The focus protection objects: Subalpine forest ecosystems, natural landscape, and key protected wild fauna and flora, such as *Rhinopithecus roxellana*, *Andrias davidianus*, *Moschus berezovskii*, *Neofelis nebulosa*, *Panthera pardus*, *Ursus thibetanus*, *Aquila chrysaetos*, *Macaca mulatta*, *Davidia* sp., *Cercidiphyllum japonicum*, *Pterostyrax psilophyllus*, etc.
Pingqian	Huangbaiqian Guogongping	The focus protection objects: Subalpine forest ecosystems, natural landscape, and key protected wild fauna and flora, such as *Moschus berezovskii*, *Neofelis nebulosa*, *Panthera pardus*, *Aquila chrysaetos*, *Syrmaticus reevesii*, *Davidia* sp., *Cercidiphyllum japonicum*, *Pterostyrax psilophyllus*, *Taxus* sp., etc.
Dongxi	Daobaying Yutuo	The focus protection objects: Subalpine forest ecosystems, natural landscape, and key protected wild fauna and flora, such as *Rhinopithecus roxellana*, *Moschus berezovskii*, *Neofelis nebulosa*, *Panthera pardus*, *Aquila chrysaetos*, *Andrias davidianus*, *Aquila chrysaetos*, *Lutra lutra*, *Davidia* sp., *Cercidiphyllum japonicum*, *Pterostyrax psilophyllus*, *Tetracentron sinense*, etc.
Laojunshan	Caiqi	The focus protection objects: Evergreen broad-leaved forest ecosystems, natural landscape, and key protected wild fauna and flora, such as *Moschus berezovskii*, *Neofelis nebulosa*, *Panthera pardus*, *Aquila chrysaeto*, *Capricornis milneedwardsii*, *Davidia* sp., *Cercidiphyllum japonicum*, *Emmenopterys henryi*, etc.
	Mucheng	The focus protection objects: Evergreen broad-leaved forest ecosystems, natural landscape, and key protected wild fauna and flora, such as *Davidia* sp., *Cercidiphyllum japonicum*, *Emmenopterys henryi*, etc.
Songziyuan	Chayuan	The focus protection objects: Evergreen broad-leaved forest ecosystems, natural landscape, and key protected wild fauna and flora, such as *Moschus berezovskii*, *Neofelis nebulosa*, *Panthera pardus*, etc.

and issuing monitoring results and recommendations in a timely manner to local governments and the public. In addition, it is responsible for timely communication with research institutions, universities, and part-time experts.

8.6.2.2 Monitoring Point Distribution

The monitoring system carries out dynamic monitoring of forests and vegetation in heritage sites, timely monitoring of the status and dynamics of forest resources, growth patterns and forest biodiversity, monitoring of forest fire prevention, invasion of harmful alien species and human activities. At the same time, the changes of animal and plant species, quantity, present situation, and habitat in heritage sites were monitored.

The management agency of the heritage has set up monitoring points in the main tourist activity space and route to monitor the number of visitors to the heritage site, tourist projects, and service facilities. On the basis of the current environmental monitoring points of hydrological and water environmental monitoring stations, more environmental monitoring points and facilities are added to monitor hydrology. The agency has established atmospheric environmental monitoring stations in the downwind direction of major tours and service areas to monitor atmospheric conditions. Also the agency has set up sound environment monitoring station near service base and main sound source to monitor sound.

8.6.2.3 The Improvement of Monitoring Methods

The monitoring of heritage sites is a combination of integrated equipment monitoring, community patrol monitoring and statistical investigation, fixed monitoring and mobile monitoring. Fixed monitoring points are set up at each site of the heritage site, and necessary equipment and professionals are equipped to use remote video surveillance system to process real-time monitoring data in the digital management system. In addition, the mobile monitoring points are established according to the monitoring objectives, and the information of the monitoring system can be analyzed and published periodically and irregularly.

8.6.2.4 Monitoring System

The monitoring systems integrate GIS, GPS, remote sensing, computers, and networks to form a heritage spatial data platform. Among them, the network video surveillance system uses a large screen system to monitor the heritage in real time, and automatically monitors all the images sent by the front end of the network, and identifies the images on the electronic map for real-time centralized monitoring. Scheduling system includes computer, network, communication tools and management information system, and decision support system which can support resource real-time scheduling.

8.6.2.5 Improvements to Equipment for Field Staff

There are plans to provide additional monitoring equipment including the provision of the following things: Equipment for fire-proofing such as fire extinguishers and alarms at monitoring and wildlife observation sites, equipment for field patrols at monitoring points (e.g., GPS), the facilities for monitoring the traffic and communication, and the equipment to monitor fire control such as extinguishers.

8.6.2.6 Monitoring Systems of Hubei Shennongjia

The current monitoring pattern cannot meet the monitoring requirements for the outstanding universal values in Hubei Shennongjia. Monitoring point number was increased, from the original six up to 20 (Table 8.4). The monitoring system included biology monitoring such as animals and their habitats, exotic species, and vegetation.

Table 8.4 Planned monitoring system of the Shennongjia

Monitoring site	Content of monitoring
Watch tower	Tourism, forest fire, environment
Guanmenshan	Biodiversity, forest fire, tourists, environment
Laojunshan	Biodiversity, forest fire, tourists, environment
Dalongtan	Biodiversity, forest fire, tourists, environment
Xiaolongtan	Biodiversity, tourism, environment
Jinhouling	Biodiversity, tourism, environment
Qianjiaping	Biodiversity, tourism, environment
Zhangbaohe	Biodiversity, forest fire, environment
Bancang	Biodiversity, forest fire, environment
Yuergou	Biodiversity, environment
Baibuti	Biodiversity, tourists, environment
Yazikou	Tourism, environment
Xiagu	Biodiversity, forest fire, environment
Bailinyan	Biodiversity, forest fire, environment
Huangboqian	Biodiversity, forest fire, environment
Guogongping	Biodiversity, forest fire, environment
Daobaying	Biodiversity, forest fire, environment
Yutuo	Biodiversity, forest fire, environment
Caiqi	Biodiversity, forest fire, environment
Mucheng	Biodiversity, forest fire, environment

Chapter 9
Tourism Development Management

The fast and convenient transportation system composed of roads and trails makes Hubei Shennongjia excellent in tour conditions. Since 1990, eco-tourism in Hubei Shennongjia has made great progress, it is now in a rapid development stage, with an annual increase of more than 20,000 tourists since 2008. In fact, the number of tourists in 2012 is 140,000 more than that in 2011. According to statistics, the number of visitors in 2013 has reached 522,000. The developing tourism in Hubei Shennongjia brings a certain pressure on the protection of natural resources and environments. During peak season, tourists are so concentrated in the scenic area that environmental capacity is saturated. As such, tourism in peak season, to a certain extent, has threatened the wild animal activity and the protection of habitat within Hubei Shennongjia. The influx of excessive tourists and vehicles, as well as the construction of tourism facilities, causes air, water, and noise pollution.

9.1 Tourism Carrying Capacity

9.1.1 Ecological Carrying Capacity

Shennongjia is located in the core water conservation area of the Middle Route Project of the South to North Water Transfer Project. Water resources are important landscape resources, as well as resources that produce negative oxygen ions in the air and drinking water, so that all domestic sewage must be treated before it is discharged. There are clear, cool, and sweet streams throughout the year in Shennongjia World Natural Heritage Site, and the water quality meets the national "Surface Water Environmental Quality Standard" (GB3838-2002) Class II. Therefore, water environment is not a limited factor of the tourism ecological carrying capacity. However, it is worth to strengthen the daily water quality monitoring to prevent sewage from entering river and drinking water sources.

Z. Xie, G. Shen, *The Outstanding Universal Value and Conservation of Hubei Shennongjia*, https://doi.org/10.1007/978-981-16-0684-7_9

Shennongjia World Natural Heritage Site is far away from industrial zone, and has a forest cover rate of 96%. The percentage of days with good air quality throughout the year reached 100%, and the air quality reached the national "Ambient Air Quality Standard" (GB3095-1996) Class I. Shennongjia World Natural Heritage Site is a typical natural forest oxygen bar, which has 166,000 negative oxygen ions per cubic centimeter of air. Therefore, air environment capacity is not a limited factor of the tourism ecological carrying capacity. Besides, as Shennongjia had clean the garbage everyday and transport the garbage out of the scenic area for treatment, the solid waste is also not a limited factor of the tourism ecological carrying capacity.

9.1.2 Spatial Carrying Capacity

The tourism spatial carrying capacity in Shennongjia should be calculated in accordance with the Code for Scenic Area Planning, for example, the reasonable density of main and general scenic spots in the area rule is 50–100 m^2/person. The area of Shennongjia World Natural Heritage Site scenic spot is 65,673 hm^2. The scenic spot opens 8 h everyday. It takes 28 h in average to complete the visit of the scenic spot. The tourism spatial carrying capacity is 328,365–1,313,460 people.

Shennongjia tourism zone is large and the distances between scenic spots are long. The measurement of environmental capacity should be divided into different scenic spots. Because of the separate scenic sites, tourists mainly travel along the scenic spot road.

1. Shennongding scenic spot: The tour paths mainly consist of the part from Shuangshenxian Yazikou to Banbiyan and the part from Shuangshenxian to Guanyindong. It also includes connection roads from the main road to each scenic site and the walking paths. There are 40,000 m incomplete tour path in Shennongding scenic spot. This scenic spot opens 8 h everyday. It takes tourists 5 h to complete all the tour paths and 3 h to get back. Therefore, the daily environmental carrying capacity of Shennongding scenic spot is calculated to be 2500–5000 people.
2. Taiziya scenic spot: The tour paths mainly consist of the part from Shuangshenxian Banbiyan to Dajialing and the part from Houzishi to Nantianmen. It also includes connection roads, walking paths, and overhead road from the main road to each scenic site. The total length of the tour path in Taiziya scenic spot is 33,000 m. It takes tourists 6 h to complete all the tour paths. The daily environmental carrying capacity of Taiziya scenic spot is calculated to be 4400–8800 people.
3. Guanmenshan scenic spot: The tour paths mainly from Shicaohekou to Guanmenshan, and also include connection roads from the main road to other scenic sites and walking paths around each scenic site. The total length of the tour path in Guanmenshan scenic spot is 30,500 m. It takes tourists 6 h to complete all

the tour paths. The daily environmental carrying capacity of Guanmenshan scenic spot is calculated to be 4066–8133 people.

4. Laojunshan scenic spot: The tour paths mainly include road along Jiuchonghe Valley, Natural bridge, connection roads to other scenic sites, and walking paths around each scenic site. The total length of the tour path in Laojunshan scenic spot is 12,000 m. It takes tourists 4 h to complete all the tour paths. The daily environmental carrying capacity of Laojunshan scenic spot is calculated to be 2400–4800 people.
5. Pingqian-Maxianping reservoir scenic spot:

   Pingqian reservoir scenic site: The tour paths mainly include road along lake, connection roads to other scenic sites, and walking paths around each scenic site. The total length of the tour path in Pingqian reservoir scenic site is 5000 m. It takes tourists 2 h to complete all the tour paths. The daily environmental capacity of Pingqian reservoir scenic site is calculated to be 2000–4000 people. Maxianping reservoir scenic site: The tour paths mainly include road along lake, connection roads to other scenic sites, and walking paths around each scenic site. The total length of the tour path in Maxianping reservoir scenic site is 5000 m. It takes tourists 2 h to complete all the tour paths. The daily environmental capacity of Maxianping reservoir scenic site is calculated to be 2000–4000 people. According to calculations, the total daily environmental capacity of the five scenic spots in Shennongjia is 17,366–34,733 people. Combining the results of the surface model and the linear-flow velocity model, the daily spatial carrying capacity of the Shennongjia Natural Heritage Site is Min (328,365–1,313,460, 17,366–34,733), which is 17,366 people.

9.1.3 Facility Carrying Capacity

In 2013, there were 727 hotels with 7337 rooms and 14,150 beds in Shennongjia tourism area. Among them, there were 13 star-rated hotels with 1658 beds, 69 non-star hotels with 5314 beds, and 645 self-employed or other type of hotels with 7178 beds. There are two tourist reception centers located in Xiaolongtan and Banbiyan in Shennongding scenic spot, with shops and restaurants. Every restaurant is able to accommodate 500 tourists at the same time. If the meal cost per person is 40 min, the number of people dining in the restaurant between 11:00 to 14:00 would be 4500. As there are 727 hotels in Shennongjia Forest District, it is completely enough for tourists to eat. There are two tourist fleets in Shennongjia. One is specially designed for foreign tourists, which has seven buses and 209 seats. The other is Shennongjia Tourism Group Corporation fleet, which has seven buses and 249 seats. In addition, there are 214 other tourist cars, including tour coaches, cars, and minibuses, with 4280 seats in total. If every car runs three times a day, the daily capacity is 14,214 people. There are ten tourist commodity producers, 123 individual local specialty shops, 27 tourist agencies in Shennongjia tourism zone, which are fully enough for tourists. Besides, the water supply, electricity supply, communication system, and transportation facilities are well set up, and not sensitive to tourist capacity. Therefore, the minimum tourism facility carrying capacity is 14,150 people.

Shennongjia has 215 tourist staff now, and the daily tourism management carrying capacity is calculated to be 21,500 people. The lengths of tour paths in the five scenic spots of Shennongjia World Natural Heritage Site are 40,000 m, 33,000 m, 30,500 m, 12,000 m, and 10,000 m, respectively. Among them, the first one, tour path of Shennongding scenic spot, is an incomplete tour path. The tourism psychological carrying capacity of the five scenic spots are 1666 people, 2933 people, 2711 people, 1600 people, and 2666 people, respectively. The tourism psychological carrying capacity of Shennongjia World Natural Heritage Site is 11,576 people.

9.1.4 Summary of Tourism Carrying Capacity

To sum up, the results of tourism carrying capacity of Shennongjia World Natural Heritage Site are as follows. The tourism spatial carrying capacity (TSCC) is 17,366 people/day. The tourism facility carrying capacity (TFCC) is 14,150 people/day. The tourism management carrying capacity (TMCC) is 21,500 people/day. The tourism psychological carrying capacity (TPCC) is 11,576 people/day. Because of the good environment in Shennongjia, ecological environment is not a limited factor of the tourism ecological carrying capacity. The best daily tourism carrying capacity of Shennongjia World Natural Heritage Site is 11,576 people. The maximum daily tourism carrying capacity is 21,500 people. The peak tourist season in Shennongjia World Natural Heritage Site is between May and October, especially in the weekends, May Day, and National Day. So, the number of days of peak tourism season, which is 180 days, is regarded to be the total tourist days. Then, the best annual tourism carrying capacity of Shennongjia World Natural Heritage Site is 2,083,680 people. The maximum annual tourism carrying capacity is 3,870,000 people.

According to the analyses of tourism carrying capacity, different parts of the tourism carrying capacity of Shennongjia World Natural Heritage Site vary greatly. The tourism spatial carrying capacity and the tourism management carrying capacity are relatively large. In view of this situation, we should find ways to improve the tourism psychological carrying capacity, so as to correspondingly improve the tourism environmental carrying capacity of the heritage site, to avoid the waste caused by investment imbalance. Shennongjia tourism zone has been open to tourists since 1990, and received 2000 tourists. It received 163,219 tourists in 2005, and 949,027 tourists from China and other countries by 2011. In 2013, it received 1,613,656 tourists, which was less than 2,083,680 people, the best tourism carrying capacity. As a result, the size of tourists in Shennongjia is still within the limit of carrying capacity. It has been reported that tourism in Shennongjia is developing rapidly, with an increasing rate of tourists of 10%. According to this rate, without any control and direct in management, the number of tourists reaches

2,148,000 in 2016, exceeding the best tourism carrying capacity, and reaches 4,185,000 in 2022, exceeding the maximum annual tourism carrying capacity.

Nowadays, Shennongjia area has been a World Natural Heritage Site, which extracts more and more tourists. In order to effectively control the number of tourists coming to Shennongjia, to achieve sustainable and healthy tourism, we should hold the eco-tourism theme, strictly protect the environment, and give full play to the features of resources. Several distinctive tourism products should be actively developed, such as ecological viewing, scientific researching, showering in the forest, health caring, mountain climbing, skiing and grass skiing, swimming, and drifting. Then we make the Shennongjia tourism products a comprehensive, multi-level, series of attractive brand, rather than the past single tourism product. The peak tourist season in Shennongjia is between May and October, which is a prominent feature of tourism. We can increase the ticket price difference between peak season and low season, stress the characteristics in low season, and monitor and evaluate the tourist size in peak season to average monthly tourist flow. At present, different scenic spots attract different number of tourists. Most tourists prefer visiting Shennongding scenic spot to Guanmenshan and Laojunshan scenic spot. To guide and disperse the crowd, we can advertise more about the less famous scenic spots, let in the tourists in batches to control the number of tourists in hotspots, and adopt a rotating rest system in important ecological scenic spots.

9.2 Influences of the Tourism Development after the Airport Opening

Shennongjia Airport was a newly opened regional airport for tourism, and Shennongjia was formally inscribed on the World Heritage List on July 17, 2016. Based on the flight data of the 2 year since the opening of Shennongjia Airport and the tourist data of Shennongjia from year 2004–2014, we analyzed the impact of the opening of Shennongjia Airport on the tourist increase of Shennongjia, firstly. And then, based on the tourist data of some of the natural World Heritage in China, we estimated the situations of tourism increase after the listing of World Heritage of Shennongjia.

According to monthly statistics of actual flight implementation data after the opening of Shennongjia Airport, it is found that the average flight implementation rate is 65.79% (minimum is 31.67%, maximum is 100%) (Fig. 9.1). The monthly average flight occupancy rate between 2014 and 2015 is 30.8% (minimum is 5.61%, maximum is 49.40%) (Fig. 9.2). The designed passenger throughput of Shennongjia Airport is 254,100 in 2020. Calculated by the average flight implementation rate, which is 65.79%, the airport passenger throughput is 161,300. The elevation of the airport is 2580 m. It located in an area of high risk of fog and thunderstorms, which seriously affect the operation of flights. Tourists' travel plans are seriously affected by the low implementation rate so that they are less willing to get aboard, resulting in

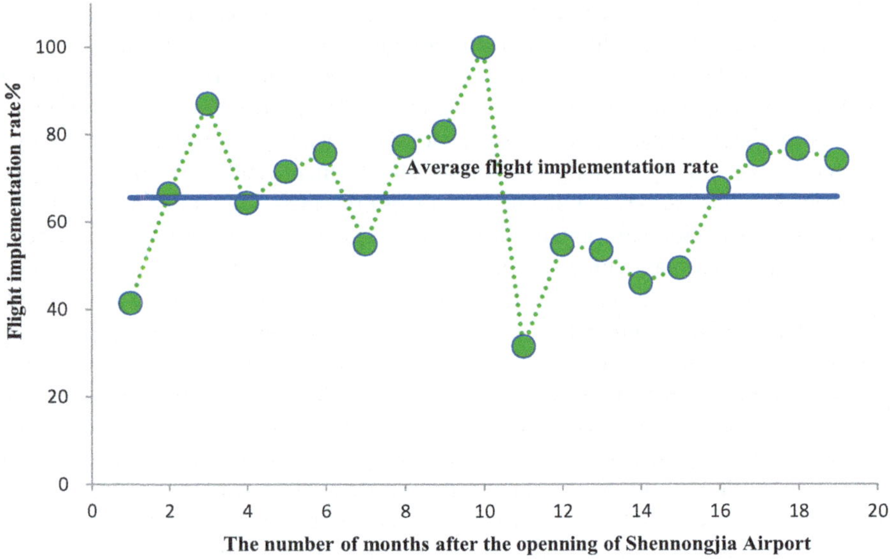

Fig. 9.1 Monthly flight implementation rate of Shennongjia Airport in 2014–2015

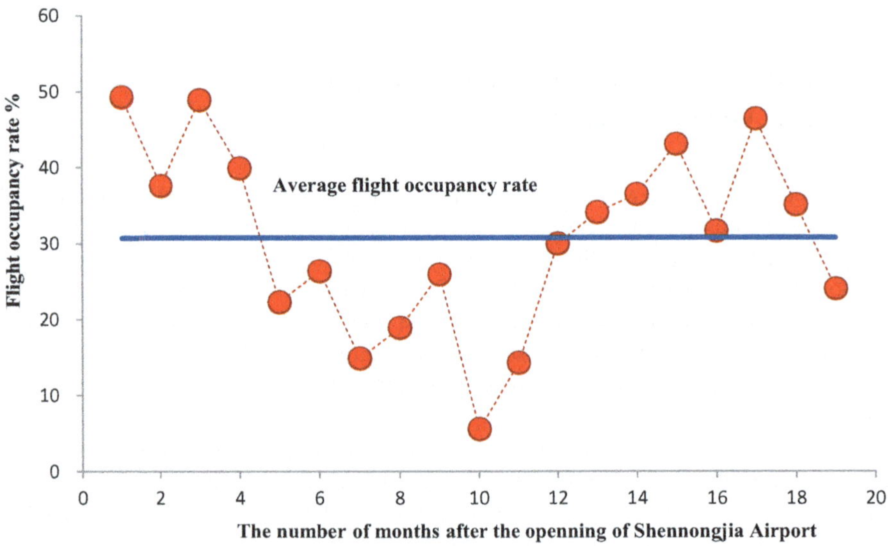

Fig. 9.2 The monthly flight occupancy rate (%) of the Shennongjia Airport from year 2014–2015

the low passenger utilization rate (monthly 30.80%). There were 20,852 passengers entering and leaving the airport in 2014, and 28,031 passengers entering and leaving the airport in 2015 (3166 transfer passengers included, so the actual number of

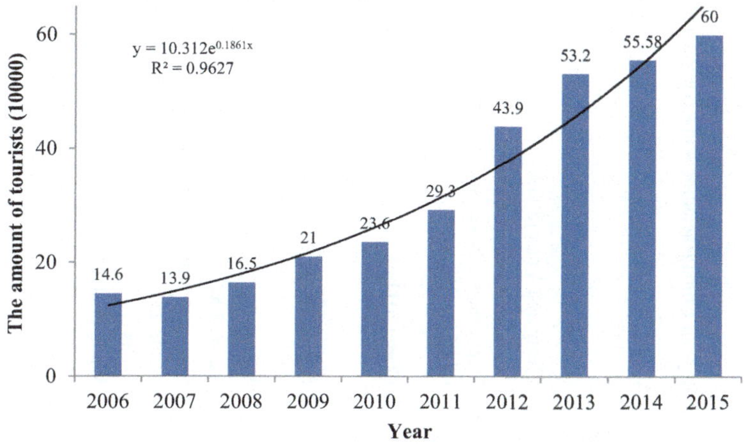

Fig. 9.3 The amount of tourists and its yearly increase rate of Shennongjia from year 2006–2015

passengers was 24,865). Between 2014 and 2015, the total number of passengers entering and leaving the airport was 45,717, with 22,900 people/year in average, only 2.06% of the total tourists (555,800 people) of Shennongjia in 2014 (Zhao et al., 2018).

The network of roads and walking paths in Shennongjia is very convenient and suitable for tourism. Shennongjia World Natural Heritage Site has developed eco-tourism since 1990, and has experienced a rapid development since 2008, with an increasing of tourists of 20,000 people per year. Especially, there were nearly 140,000 more tourists in 2012 than in 2011 (Fig. 9.3). According to the trend line in the figure, the yearly increase rate of tourists in Shennongjia was 19.44% between 2006 and 2014 (Zhao et al., 2018). The annual average increasing rate of tourists of natural World Heritage in China (including some natural and cultural World Heritage) (Table 9.1) shows that after the heritage been listed as a World Heritage, tourists increasing rate is the lowest in Huangshan (6.42%), the highest in Wulong (41.99%), and 20.24%±13.78% in average.

Calculated by the tourist increasing rate in Shennongjia in recent years, and annual average increasing rate of tourists of natural World Heritage after been listed as a World Heritage in China (including some natural and cultural World Heritage), the tourist increasing rate in Shennongjia is 20.64%±13.74% (i.e., 6.90–34.38%) (Zhao et al., 2018).

Table 9.1 Annual average increasing rate of tourists after the World Heritage listing of part of natural world heritage in China (including some natural and cultural World Heritage)

World heritage	The year of listing of world heritage	The amount of tourists (10,000)	Annual average increasing rate of tourists	Statistical time	Number of years	Literature
Mount Tai (natural and cultural world heritage)	1987	280–1011	8.94	1990–2005	16	Zheng and Yuan (2009)
Mount Huang (natural and cultural world heritage)	1990	67–160.19	6.42	1990–2004	15	Huangshan City statistical yearbook (2005)
Mount Wuyi (natural and cultural world heritage)	1999	193–874	12.32	1999–2012	14	Chen (2011)
Mount Emei (natural and cultural world heritage)	1999	60–330	12.94	1998–2010	13	Chen (2011)
Jiuzhaigou	1992	15.94–221.01	24.50	1994–2006	13	Yan et al. (2009)
Wulingyuan	1992	165–1670	17.98	1992–2006	15	Yang (2008)
Wulong	2007	164–1908	41.99	2007–2014	8	Statistical bulletin of the national economic and social development of Wulong County
Mount Sanqingshan	2008	150–807.71	40.03	2008–2013	6	Wang et al. (2014)
Annual growth rate			20.64			

9.3 The Responses and Strategies for Managing Tourist Development

Based on an analysis of tourist numbers of world natural heritage sites in China, the annual increase rate of tourists after listing is 14%. Tourists to the Shennongjia property were 555,800 in 2015. The tourism environmental capacity is 1.83 million per year, as estimated in the Management Plan for Hubei Shennongjia. Thus, annual tourists are not expected to exceed the tourism environmental capacity.

9.3.1 The Responses of the State Party China

The State Party China plans to strengthen policies and institutions, increase staffing and resources, integrate Hubei Shennongjia into the management framework of China's national park system. The State Party China divert visitors to West Hubei Eco-cultural Tourism Circle from Hubei Shennongjia by strengthening the radiation roles of the Circle, and promote the franchise tourism and improve the management efficiency of eco-tourism to reduce environmental pressure from increased tourism. To meet the demand of tourism increases, 22 employees qualified in tourism management, environmental conservation, wildlife conservation and nature reserve management, forest conservation, etc., are employed, which make the proportion of the staff for the tourism management and protection increase from 26.4% in 2018 to 35.1% in 2020.

9.3.2 Strategies for Managing Tourist Numbers

The daily visitor number was strictly controlled at the ticket office and ticket checking ports during tourism peak periods (tourist season), by a smart ticketing system and real-time statistics system on the quantity of visitors and vehicles. Implementing the online ticket booking system and forecasting the number of the daily booked tourists ensure that visitors can adjust their travel schedule in time. Immediate broadcasting of the visitor flow and traffic flow in every scenery area by outdoor websites, LED, WeChat, Radio, and so on guide visitors' traveling plans. At the Muyu New Town, located outside the buffer zones, catering, accommodation, as well as green vehicles, were provided in adequate supply to relieve the stress from tourism growth. Measures were taken to improve tourism service quality, such as optimizing the layout of scenic spots and eco-tourism trails, enhancing the ability for travel consultation and medical assistance, strengthening the team construction and skills training, increasing tourism logos and environmental protection facilities, establishing specialized environmental management agencies and systems, promoting the implementation of ISO14000 standards, collecting garbage and sewage in a timely manner, and transporting them outside Hubei Shennongjia for appropriate disposal. Only green vehicles are allowed and other vehicles are prohibited from entering Hubei Shennongjia and its buffer zone, promoting traffic safety and reducing environmental pollution.

9.3.3 Countermeasure and Suggestion

According to the analyses of daily flight implementation data between 2014 and 2015, the flight implementation rate and the flight occupancy rate were very low. Tourists' travel plans are seriously affected by the low implementation rate so that they are less willing to get aboard, resulting in the low passenger utilization rate. Calculated by the current flight implementation rate, the airport passenger throughput is only 161,300. Between 2014 and 2015, the average number of passengers entering and leaving the airport was 22,900 people/year, only 2.06% of the total tourists (555,800 people) of Shennongjia in 2014. It shows that the opening of Shennongjia Airport contributes little to the increasing of tourists.

According to the tourist increasing rate in Shennongjia in recent years and annual average increasing rate of tourists of natural World Heritage after been listed as a World Heritage in China (including some natural and cultural World Heritage), the number of tourists exceeds tourism carrying capacity around 2020. Therefore, it is necessary to take actions to cope with the possible increasing of tourists. To control the number of tourists and improve the quality and efficiency of management, it is necessary to protect the sustainable of resources and ensure the sustainable development of tourism. To cope with the possible increasing of tourists and the threat of exceeding tourism carrying capacity, it is necessary to strengthen rules and policy, increase staff and facilities, improve quality of management staff and efficiency of management, and use information technology tools to control the number of tourists and guide them, so that to relieve the pressure of tourism increasing.

It is suggested that strengthen rules and policy and increase staff and facilities are necessary to cope with and relieve the pressure of tourism increasing, for example, put Shennongjia under the management of national park system to distract the pressure of tourism, strengthen the radiation function of west Hubei eco-tourism circle to divert the number of tourists in Shennongjia, carry out the franchise model of management in Shennongjia to improve efficiency of tourism management, and increase headcount of management and service staff and so on. It is suggested that active and effective control measures should be taken to guide tourists and cope with the pressure of tourism, for example, use intelligent ticketing system, passenger flow, and car flow statistic system to strictly control the amount of tourists during peak hours, implement online ticket booking system, and predict daily booking number in time to ensure that tourists can adjust their schedules according to the tourist flow of the scenic site, use websites, LED, WeChat, podcast, TV, and other tools to report the real-time traffic flow and tourist flow of the scenic site to guide the diversion of tourists and so on.

It is suggested that the reception ability, facilities, service quality, and environmental protection in the scenic site should be improved, for example, build more restaurants and hotels, equip more environmentally friendly vehicles in Muyu new town, the area outside buffer zone, to improve the reception ability outside buffer zone, optimize the layout of viewing platform, beautify the eco-tourism walking path, improve consultation and medical assistance ability, enhance the development and training of management and service team, and improve staff and service quality. Besides, we should add tourist signals and environmental protection facilities, establish specialized environmental management institutions and policies, and collect, transfer, and dispose garbage and sewage in time. We should also provide environmental friendly buses in the scenic site and prevent the entry of outside cars to ensure driving safety and reduce environmental pollution caused by exhaust gas emission.

Chapter 10
Integrated Management and Community Development

10.1 Integrated Management

To ensure operation of Hubei Shennongjia, the government of Hubei Province consulted local governments, monitoring organizations, universities and research institutes, and then established the Leading Group for Application for World Heritage of Hubei Province. To further strengthen the management of Hubei Shennongjia and its buffer zone, the State Party China has authorized the Administrative Bureau of Shennongjia to establish the Co-management Committee of the Property of Shennongjia-Badong (hereafter referred to as the Co-management Committee) in January 2016. The Co-management Committee is responsible for the routine management and multilateral coordination of Hubei Shennongjia under the supervision of MOHURD.

10.1.1 The Establishment of the Co-Management Committee

The Co-management Committee consists of one director, one vice-director, and several committee members. The director of the Administrative is appointed as the director of the committee, and the director of the Administrative Bureau of Badong Provincial Nature Reserve for Golden Monkey as the vice-director of the committee. Other members of the Committee are elected from the two nature reserve administrative bureaus. To strengthen the comprehensive management of the property, the State Party China established the Integrated Committee to substitute the Co-management Committee of the Property of Shennongjia-Badong (Fig. 10.1).

The Integrated Committee, supervised by MOHURD is jointly established by Shennongjia Forest District Government and Enshi Tujiazu Miaozu Autonomous Prefecture. It is responsible for the integrated protection of heritage among multilateral departments. The committee consists of the Administrative Bureau of Hubei

Z. Xie, G. Shen, *The Outstanding Universal Value and Conservation of Hubei Shennongjia*, https://doi.org/10.1007/978-981-16-0684-7_10

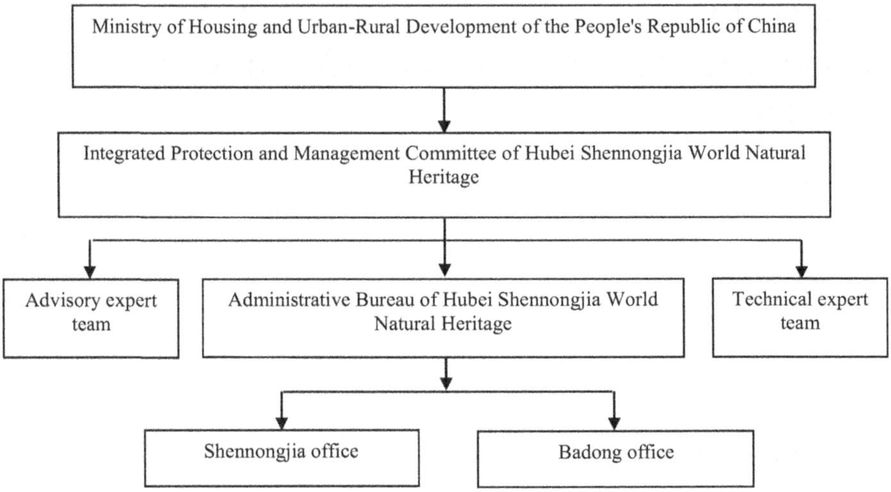

Fig. 10.1 Organization of the Committee

Shennongjia World Natural Heritage, advisory expert team, and technical expert team. The Administrative Bureau of Hubei Shennongjia World Natural Heritage, including Shennongjia office and Badong office, is responsible for the routine management of heritage. The position of director of the committee is held by the leader of the province government. The two vice-directors held by leaders from the local government, as well as six members elected from Shennongjia office and Badong office. The aim of the Integrated Committee is to develop a long-term working mechanism for the management of Hubei Shennongjia and its buffer zone including the corridors and the stepping stones. The responsibilities of the committee are to: (1) carry out environmental monitoring, scientific research and demonstration; (2) strengthen scientific protection and management; (3) fortify law enforcement; (4) develop plans and regulations; and (5) set up communication and coordination models.

10.1.2 Co-Management with the Community

An integrated mechanism of heritage site and community is vital to heritage management. The Integrated Committee had established a harmonious co-protection and joint-management mechanism with local communities. Participatory involvement by rural communities promotes the integrity of heritage protection. The Integrated Committee limited the residents' activities in Hubei Shennongjia by forbidding logging, hunting, and digging for medicine. Meanwhile, the committee encourages the residents to emigrate Hubei Shennongjia. Also, the committee promotes and popularizes the outstanding universal value of the heritage to the community, so that the community can realize the importance of its protection. The Integrated Committee educates and trains the community to understand the specifications and technologies of heritage protection. Thus, the community participates actively in the administrative affairs of the heritage protection, which ensures the effectiveness and efficiency of the heritage monitoring and protection.

The Integrated Committee helps the community develop by joint management of community businesses. The Integrated Committee helps the community improve their production mode and life style by alleviating poverty, introducing new technologies and products, substituting electricity and energy-efficient stoves for fuel wood. Conforming to principles of voluntary, democratic, equality, and mutual benefits, the Integrated Committee and the local community share the profits in implementing these projects, which not only increases the incomes of the community, but also prevents ecological destruction from community development. The mechanism of co-protection and joint management ensures the integrity and sustainability of the outstanding universal value of Hubei Shennongjia.

10.1.3 Information Platform for Integrated Management

The State Party China plans to establish an integrated management information platform for Hubei Shennongjia. Using information technology such as wireless broadband communication, intelligent sensors, high resolution earth observation, satellite positioning, "cloud" computing and "big data," the platform integrated heritage management (including OUV protection and monitoring, disaster prevention and control), and local community development into the information platform to promote comprehensive management of the heritage.

Firstly, wireless network base stations were deployed in the heritage site and its buffers to ensure that the network covers key monitoring areas, major patrol routes, and protection stations. Secondly, video capture devices, infrared cameras, habitat factor sensors, and fire hazard sensors were set up in the corridors and stepping stones to monitor the ongoing ecological processes of the subtropical mixed evergreen and deciduous broad-leaved forests and the mountain altitudinal biological zones, and to monitor the distribution and population dynamics of relic, rare, endangered, endemic, and type species, as well as deciduous woody species. In addition, the video capture devices, infrared cameras, habitat factor sensors, and fire hazard sensors were installed in the buffer zones to monitor human activities and the dynamics of forest fires and pests, which guarantee early warning of human disturbance, as well as forest fires and pests. Equipped with handheld multifunctional intelligent devices, the administrators, inspectors, and law enforcement officers from the Integrated Committee can record events that occur in the area, then send the information to multimedia command dispatching systems. Thus, the Integrated Committee can command and dispatch the staff and resources to deal with the events in real time.

10.2 Willingness of the Community to Property Conservation

Improving the production technology and production levels of community residents living in, or close to protected areas, can be an effective measure for environmental protection. Combining the development of Hubei Shennongjia with community development can help relieve the stress that the community places on the natural resources of Hubei Shennongjia. Whether and how local community inhabitants participate in affect the protection and exploitation of natural heritage directly. Here, we made the questionnaire survey for the willingness investigation of the community involvement in Hubei Shennongjia in 2014 based on 302 questionnaires. We investigated the local permanent residents for the survey, including farmers, individual and commercial households, enterprise employees and Reserve staff, etc. (Table 10.1).

10.2.1 Attitude of Communities to Hubei Shennongjia Protection

According to the survey, all the respondents think that it is important to protect Hubei Shennongjia, and 98% of the respondents think that it is necessary to make the Shennongjia to be the world natural heritage, 69% of the respondents believe that the local resident activities have no harm to Hubei Shennongjia (Table 10.2). And at the same time, 69% of the respondents believe that protection of Hubei Shennongjia has no effects on their livings, and 98% of the respondents have the confidence for the protection of Hubei Shennongjia (Xiong et al., 2017). More than 78% of the respondents believe that they can establish a close rapport with the local governor for Hubei Shennongjia protection (Table 10.2). More than 85% of the respondents believe that the environment will be better when Hubei Shennongjia is protected. And more than 81% of respondents believe that developments in Hubei Shennongjia have no negative effects on the environment protection. Ninety-five percents of the respondents are willing to give their supports for the protection of Hubei Shennongjia. Among which, 45% of the respondents can supply the labors and 23% of the respondents can give financial supports, and 17% of respondents can help the local governors to manage Hubei Shennongjia (Xiong et al., 2017). Eighty-six percents of the respondents think that the local governments and enterprises should be responsible for the environment protection of Hubei Shennongjia (Table 10.2) (Xiong et al., 2017).

Table 10.1 Respondents characteristics of community participation in our study

Age structure		Sex		Education					Income			
Age	Percentage	Male	Female	Below primary school	Primary school	Junior middle school	Senior middle school	Above junior college	<1000	1000–3000	3000–5000	>5000
20–29	7.4	52.4	47.6	4.8	9.5	47.6	28.6	9.5	19.0	81.0	0.0	0.0
30–39	22.1	76.2	23.8	6.5	19.4	46.7	22.6	4.8	27.0	66.6	4.8	1.6
40–49	35.1	82.0	18.0	13.1	32.3	38.4	14.1	2.1	29.3	64.6	6.1	0.0
50–59	24.9	87.3	12.7	16.7	38.9	31.9	12.5	0.0	30.6	63.8	2.8	2.8
>60	10.5	93.3	6.7	29.0	35.5	19.4	12.9	3.2	64.5	35.5	0.0	0.0

Table 10.2 Community awareness in the protection and exploitation of natural heritage

Item	Results			
Importance of heritage protection	65.8 very important	34.2 important	0.0 not important	0.0 no awareness
Destructive activity to heritage	6.0 very serious	5.6 serious	69.2 not serious	19.2 no awareness
Influence on production and life	5.4 very serious	4.4 serious	69.4 not serious	20.8 no awareness
Compatibility between protection and exploitation	78.0 good	5.3 bad	16.7 no awareness	
Influence on environment caused by protection	85.2 positive	4.9 negative	9.9 no awareness	
Influence on environment caused by exploitation	81.6 positive	7.6 negative	10.8 no awareness	

10.2.2 Community Willingness to Property Protection

The survey results show that more than 89% of the respondents hope to have a hand in planning and making decision for the development of Hubei Shennongjia if the local governments and enterprises permit (Table 10.3). And more than 64% of the respondents indicates that the planning and decision-making should consider the interests of the local communities. The survey results show that more than 86% of respondents believe that they expect to participate in the specific business managements if the local government provides supports, the enterprises like to cooperate, and local communities help organization (Xiong et al., 2017). And among which more than 41 of the respondents prefer to engage in the catering and accommodation, and more than 24% of the respondents prefer to be employed by the enterprises developed for Hubei Shennongjia (Table 10.3).

The survey results show that more than 69% of the respondents believe that the incomes have increased greatly since the establishment of the State Nature Reserve of the Shennongjia due to the involvement in the food, beverage, and lodge and transportation industry (Table 10.3). More than 81% of the respondents are satisfied with the income and the management environments. More than 63% of the local residents hope the local governments to participate in the benefits allocation. More than 34% of the local residents expect that the local governments, enterprises, and local communities guide the benefit allocation. And more than 73% of the respondents think that the local governments should make rules such as state subsidy, enterprise compensation, and enterprise profit-sharing system for maintaining the stability of community interest distribution. More than 60% of the respondents hope to carry joint business operation for the development of Hubei Shennongjia. More than 41% of the respondents hope to engage in the catering and accommodation. The survey results show that more than 75% of the respondents believe that it is the government that can provide strong security for the local community participation. And more than 97% of the respondents believe that the government can develop

Table 10.3 Community participation status in the protection and exploitation of natural heritage

Item	Results				
Whether participated in planning decisions	89.4 yes	2.3 no	8.3 no awareness		
Whether planning decisions considered your benefit	23.6 extremely	41.2 rather	22.6 hardly	12.6 no awareness	
Attitude to applying for world natural heritage	98.3 support	0.0 oppose	1.7 not concerned		
Whether participated in business management	86.9 yes	13.1 no			
What business participated in	24.8 employment	41.0 accommodation and catering	15.0 souvenir	19.2 transport	
Whether be pleased with present business operation	13.2 very satisfactory	68.3 satisfactory	18.5 not satisfactory		
Whether improved your income from exploitation	15.3 greatly improved	54.5 improved	6.6 not improved	0.0 decreased	23.6 no awareness
What business operation improved your income	25.0 employment	40.2 accommodation and catering	14.5 souvenir	20.3 transport	
Whether be pleased with present exploitation	12.7 very satisfactory	70.2 satisfactory	17.1 not satisfactory		
Who insured you participating in	75.6 government	6.2 enterprise	9.4 government and enterprise	4.6 community	4.2 no awareness
What method insured you participating in	81.2 policy	3.9 management	4.9 skill	0.6 social relation	9.4 no awareness
Your faith in social security	35.4 full hearted	62.3 hearted	0.0 not hearted	2.3 no awareness	

measures for securing the local community participation (Table 10.4) (Xiong et al., 2017).

10.3 Community Development

The community had established a good relationship between the property's staff and community residents, in terms of nature conservation, and community economic development and public welfare. The community had formed an effective system of

Table 10.4 Willingness and appeal of community participation in the World Heritage

Item	Results (%)				
Faith in heritage protection	39.9 full hearted	58.4 hearted	0.0 not hearted	1.7 no awareness	
To publicize for heritage protection	39.0 very willing	60.0 willing	0.3 not willing	0.7 no awareness	
To participate in training	28.3 very willing	67.7 willing	1.7 not willing	2.3 no awareness	
To be employed for heritage protection	39.3 very willing	57.0 willing	1.4 not willing	2.3 no awareness	
To participate in decision management	26.6 very willing	54.8 willing	1.7 not willing	16.9 no awareness	
Whether support for protection and exploitation	95.0 provide	b 1.7 not provide	3.3 not concerned		
Your predominance to participate in	23.0 investment	14.2 folk art	45.1 labor force	17.7 management	
Who should provide investment	65.6 government	7.6 enterprise	13.4 govern-ment and enterprise	4.5 others	8.9 no awareness
What management you will participate in	44.1 indi-vidual operation	55.9 coopera-tive operation			
Your most likely business operation	23.9 employment	41.7 catering	16.3 souvenir	18.1 transport	
Who conduct the profit sharing	63.8 government	1.0 enterprise	0.7 community	34.5 all of the three	
Major profit-sharing source	64.3 gov-ernment subsidy	3.0 enterprise compensation	26.3 busi-ness management	6.4 corpo-rate dividends	

community condominium between the property and the surrounding communities, which ensured that Community residents actively participated in the management and optimum utilization of resources of Hubei Shennongjia. Through community condominium and development, community achieved a good balance between protection and development, effective protection and sustainable use of the property's natural resources. Ultimately, community completed the condominium and development of between Hubei Shennongjia and the surrounding communities.

In recent years, Hubei Shennongjia has carried out various activities to help enhance community residents' lives and methods of production. As a result, more and more community residents support and understand the management and construction of Hubei Shennongjia. This is an important way for nature conservation to be increased. In order to win community residents' support for environmental protection, improve their standards of living and quality of life, and reduce the pressure they place on natural resources, a plan had been carried out to construct 600 biogas digesters.

The average income of the residents in Hubei Shennongjia is lower than those of residents in the surrounding communities. This is partly due to limitations on their activities caused by natural resource protection, such as the fact that they are forbidden from logging, hunting, and digging for medicine. Community residents have made tremendous sacrifices for conservation of natural resources. Because of their sacrifices, Shennongjia has managed to retain special landscapes and eco-tourism features. In order to allow the community residents to benefit from eco-tourism, residents should be encouraged to establish eco-tourism service facilities. The property allows community residents to share in the benefits of conservation in order to enhance community residents' participation in protection and publicity. A plan had been carried out to aid community residents to build 20 eco-tourism farms and ten eco-tourism souvenir development projects.

There are nearly 7000 residents within Hubei Shennongjia. Their basic production and living is derived from Hubei Shennongjia. They threaten the natural resources of Hubei Shennongjia. Because of the area's poor infrastructure, it is difficult for these residents to get good education for their children, medical service, or economic development. These residents can therefore be invited to migrate outside of the core area of Hubei Shennongjia to improve their own lives and to help effectively protect natural resources and wildlife habitats. In order to reduce resource consumption and disturbance and strengthen the protection of natural resources in Hubei Shennongjia, while also aiding poorer residents, a plan had been carried out to gradually migrate residents outside of Hubei Shennongjia. Strengthening the implementation of the national policy of "Returning Farmland to Forest," key ecological projects of "Natural Forest Protection Project" and "Yangtze River Protection Forest Engineering," gradually reduce the dryland area in Hubei Shennongjia. The land for orchard, tea industry, and tourism facilities should be strictly controlled. Rural residential land should be gradually reduced in Hubei Shennongjia by the ecological immigration.

References

Bahram, M., Polme, S., Koljalg, U., Zarre, S., & Tedersoo, L. (2012). Regional and local patterns of ectomycorrhizal fungal diversity and community structure along an altitudinal gradient in the Hyrcanian forests of northern Iran. *New Phytologist, 193*, 465–473.

Bai, X., Ling, W. L., Duan, R. C., Qiu, X. F., Liu, C. X., Kuang, H., Gao, Y. J., Zhou, L., Chen, Z. W., & Lu, S. S. (2011). Mesoproterozoic to Paleozoic Nd isotope stratigraphy of the South China continental nucleus and its geological significance. *Science China-Earth Sciences, 11*, 1665–1674.

Ban, J. D., & Qi, G. S. (2001). *A study of vegetation of western Hubei*. Huazhong University of Science and Technology Press (In Chinese).

Beckmann, J. P., Clevenger, A. P., Huijser, M. P., & Hilty, J. A. (2010). *Safe passages: Highways, wildlife, and habitat connectivity*. Island Press.

Cai, B., Li, J. T., Chen, Y. Y., & Wang, Y. Z. (2016). Exploring the status and causes of China's threatened reptiles through the red list assessment. *Biodiversity Science, 24*, 578–587. (In Chinese).

Chen, Y. (2011). *Study on sustainable development of World Heritage Mountain Wuyi*. MsD dissertation, Beijing: China University of Geosciences.

Chen, W. W., Xie, Z. Q., & Zhou, Y. B. (2019). Proximity to roads reduces acorn dispersal effectiveness by rodents: Implication for forest regeneration and management. *Forest Ecology and Management, 433*, 625–632.

Chetkiewicz, C. L. B., St. Clair, C. C., & Boyce, M. S. (2006). Corridors for conservation: Integrating pattern and process. *Annual Review of Ecology, Evolution, and Systematics, 37*, 317–342.

CITES. (2014). https://www.iucnredlist.org/search?query=2014&searchType=species

CITES. (2016). https://cites.org/sites/default/files/notif/E-Notif-2016-043.pdf

Deng, S. Y., Dong, X. Z., Ma, M. Z., Zang, Z. H., Xu, W. T., Zhao, C. M., Xie, Z. Q., & Shen, G. Z. (2018). Evaluating the effectiveness of Shennongjia national nature reserve based on the dynamics of forest carbon pools. *Biodiversity Science, 26*(1), 27–35. (In Chinese).

Editorial Committee of Vegetation Map of China, Chinese Academy of Sciences. (2007). *Vegetation map of People's Republic of China*. Geological Publishing House (In Chinese).

Fan, D. Y., Gao, X. M., Du, Y. J., Shen, G. Z., Xu, W. T., Xiong, G. M., Zhao, C. M., Zhou, Y. B., & Xie, Z. Q. (2017a). Diversity and representativeness of deciduous woody plants in Shennongjia world natural heritage site, China. *Biodiversity Science, 25*(5), 498–503. (In Chinese).

Z. Xie, G. Shen, *The Outstanding Universal Value and Conservation of Hubei Shennongjia*, https://doi.org/10.1007/978-981-16-0684-7

Fan, D. Y., Gao, X. M., Yang, Y., Xiong, G. M., Shen, G. Z., Zhao, C. M., Xu, W. T., Zhou, Y. B., & Xie, Z. Q. (2017b). Archaic nature of seed plants from the Shennongjia world natural heritage site, China. *Plant Science Journal, 35*(6), 835–843. (In Chinese).

Fei, L., & Ye, C. Y. (1982). Geographical distribution characteristics of amphibian in Hubei Province, including a new species. *Acta Zoologica Sinica, 3*, 293–301.

Fei, L., Ye, C. Y., & Jiang, J. P. (2012). *Colored atlas of Chinese amphibians and their distribution*. Sichuan Science and Technology Press (In Chinese).

Gao, X. M., Xu, W. T., & Xie, Z. Q. (2019). *Plant catalogue of Shennongjia*. Science Press (In Chinese).

Gaston, K. J., Jackson, S. F., Cantú-Salazar, L., & Cruz-Piñón, G. (2008). The ecological performance of protected areas. *Annual Review of Ecology Evolution, and Systematics, 39*, 93–113.

Ge, J. L., Xiong, G. M., Deng, L. Q., Zhao, C. M., Shen, G. Z., & Xie, Z. Q. (2012). Community dynamics of a montane *Fagus engleriana–Cyclobalanopsis* multiervis mixed forest in Shennongjia, Hubei, China. *Biodiversity Science, 20*(6), 643–653. (In Chinese).

Ge, J. L., Xiong, G. M., Zhao, C. M., Shen, G. Z., & Xie, Z. Q. (2013). Short-term dynamic shifts in woody plants in a montane mixed evergreen and deciduous broadleaved forest in central China. *Forest Ecology and Management, 310*, 740–746.

Gomez-Hernandez, M., Williams-Linera, G., Guevara, R., & Lodge, D. J. (2012). Patterns of macromycete community assemblage along an elevation gradient: Options for fungal gradient and metacommunity analyses. *Biodiversity and Conservation, 21*, 2247–2268.

Grassman, L. I., Jr., Tewes, M. E., & Silvy, N. J. (2005). Ranging, habitat use and activity patterns of binturong Arctictis binturong and yellow-throated marten Martes flavigula in north-central Thailand. *Wildlife Biology, 11*, 49–57.

Hanski, I., & Ovaskainen, O. (2000). The metapopulation capacity of a fragmented landscape. *Nature, 404*, 755–758.

He, F. (2006). Forest vegetation in Japan. *Journal of Sichuan Forestry Science and Technology, 3*, 38–41.

Hill, M. O., Carey, P. D., & Eversham, B. C. (1993). T*he role of corridors, stepping stones and islands for species conservation in a changing climate*. English Nature.

Hodkinson, I. D. (2005). Terrestrial insects along elevation gradients: Species and community responses to altitude. *Biology Review, 80*, 489–513.

Huang, J., & Xia, Z. H. (2013). Research on the change characteristics for Shennongjia forest vegetation index based on MODIS-EVI data. *Meteorological and Environmental Sciences, 3*, 39–43. (In Chinese).

Huangshan City Bureau of Statistics. (2005). Huangshan City statistical yearbook, Huangshan. https://cites.org/sites/default/files/notif/E-Notif-2014-053.pdf

IUCN/SSC. (2011). Habitats Classification Scheme, Version 3.0. http:///www.Iucnredlist.org/technical-documents/classificationschemes/habitats-classification-scheme-ver3.

IUCN Red List. (2014). https://www.iucnredlist.org/search?query=2014&searchType=species

IUCN Red List. (2016). https://www.iucnredlist.org/search?query=2016&searchType=species

Jiang, Z. (2015). *China's mammals diversity and geographic distribution*. Science Press (In Chinese).

Jiang, Z., Liu, F., Zhang, W., Luan, X. F., Li, D. Q., & Li, J. (2015). Changes in land cover in Shennongjia forest region in the past 3 decades based on interpretation of landsat images. *Journal of Yangtze River Scientific Research Institute, 33*(6), 150–154. (In Chinese).

Jiang, J. P., Xie, F., Zang, C. X., Cai, L., Li, C., Wang, B., Li, J. T., Wang, J., Hu, J. H., Wang, Y., & Liu, J. Y. (2016a). Assessing the threat status of amphibians in China. *Biodiversity Science, 24*, 588–597. (In Chinese).

Jiang, Z. G., Jiang, J. P., Wang, Y. Z., Zhang, E., Zhang, Y. Y., Li, L. L., Xie, F., Cai, B., Cao, L., Zheng, G. M., Dong, L., Zhang, Z. W., Ding, P., Luo, Z. H., Ding, C. Q., Ma, Z. J., Tang, S. H., Cao, W. X., Li, C. W., . . ., Ping, X. G. (2016b). Red list of China's vertebrates. *Biodiversity Science*, 24, 500–551. (In Chinese).

Jiang, Z. G., Li, L. L., Luo, Z. H., Tang, S. H., Li, C. W., Hu, H. J., Ma, Y., Wu, Y., Wang, Y. X., Zhou, K. Y., Liu, S. Y., Feng, Z. J., Cai, C., Zang, C. X., Zeng, Y., Meng, Z. B., Ping, X. G., & Fang, H. X. (2016c). Evaluating the status of China's mammals and analyzing their causes of endangerment through the red list assessment. *Biodiversity Science, 24*, 552–567. (In Chinese).

Joppa, L. N., & Pfaff, A. (2010). Global protected area impacts. *Proceedings of the Royal Society B: Biological Sciences, 278*, 1633–1638.

Korner, C. (2007). The use of "altitude" in ecological research. *Trends in Ecology and Evolution, 22*, 569–574.

Li, H. L. (1952). Floristic relationships between Eastern Asia and Eastern North America. *Transactions of the American Philosophical Society New Series, 2*, 371–429.

Li, Y. M. (2007). Terrestriality and tree stratum use in a group of Sichuan snub-nosed monkeys. *Primates, 48*, 197–207.

Li, Y. M., Xu, L., Ma, Y., Yang, J. Y., & Yang, Y. H. (2003). The species richness of nonvolant mammals in Shennongjia Nature Reserve, Hubei Province, China: Distribution patterns along elevation gradient. *Biodiversity Science, 1*, 1–9. (In Chinese).

Liu, J. Y., Xie, Z. Q., Shen, G. Z., Fan, D. Y., Xiong, G. M., Zhao, C. M., Zhou, Y. B., & Xu, W. T. (2018). Dynamics and analysis of vegetation fraction changes in Shennongjia forest district during 1998 to 2013 by using SPOT-VEGETATION NDVI data. *Acta Ecologica Sinica, 38* (11), 3961–3969. (In Chinese).

Ma, M. Z., Shen, G. Z., Xiong, G. M., Zhao, C. M., Xu, W. T., Zhou, Y. B., & Xie, Z. Q. (2017). Characteristic and representativeness of the vertical vegetation zonation along the altitudinal gradient in Shennongjia natural heritage. *Chinese Journal of Plant Ecology, 41*(11), 1127–1139. (In Chinese).

Malhi, Y., Silman, M., Salinas, N., Bush, M., Meir, P., & Saatchi, S. (2010). Introduction: Elevation gradients in the tropics: Laboratories for ecosystem ecology and global change research. *Global Change Biology, 16*, 3171–3175.

McCain, C. M. (2009). Global analysis of bird elevational diversity. *Global Ecology and Biogeography, 18*, 346–360.

McGuire, J. L., Lawler, J. J., McRae, B. H., Nunez, T. A., & Theobald, D. M. (2016). Achieving climate connectivity in a fragmented landscape. *Proceedings of the National Academy of Sciences United States of America, 113*, 7195–7200.

Nakashizuka, T., & Matsumoto, Y. (2002). *Diversity and interaction in a temperate forest community: Ogawa Forest Reserve of Japan*. Springer.

Ohsawa, M. (1984). Differentiation of vegetation zones and species strategies in the subalpine region of Mt. Fuji. *Vegetation, 57*, 15–52.

Olson, D. M. (1994). The distribution of leaf litter invertebrates along a Neotropical altitudinal gradient. *Journal of Tropical Ecology, 10*, 129–150.

Ouyang, Z. Y., Zheng, H., Xiao, Y., Polasky, S., Liu, J. G., Xu, W. H., Wang, Q., Zhang, L., Xiao, Y., Rao, E. M., Jiang, L., Lu, F., Wang, X. K., Yang, G. B., Gong, S. H., Wu, B. F., Zeng, Y., Yang, W., & Daily, G. C. (2016). Improvements in ecosystem services from investments in natural capital. *Science, 352*, 1455–1459.

Palin, O. F., Eggleton, P., Malhi, Y., Girardin, C., Rozas-Davila, A., & Parr, C. (2011). Termite diversity along an Amazon-Andes elevation gradient, Peru. *Biotropica, 43*, 100–107.

Pauleit, S., Ennos, R., & Golding, Y. (2005). Modelling the environmental impacts of urban land use and land cover change-a study in Merseyside, UK. *Landscape and Urban Planning, 71*, 295–310.

Pérez-Hernández, C. G., Vergara, P. M., Saura, S., & Hernández, J. (2015). Do corridors promote connectivity for bird-dispersed trees? The case of Persea lingue in Chilean fragmented landscapes. *Landscape Ecology, 30*, 77–90.

Qi, C. J., Yu, X. L., Zhen, Z., & Yin, G. P. (1998). Notes on endemic seed plants in the floristic region of central China. *Journal of Central South Forestry University, 18*, 1–4. (In Chinese).

Qian, H. (2001). A comparison of generic endemism of vascular plants between east Asia and north America. *International Journal of Plant Sciences, 162*, 191–199.

Qin, H. N., Yang, Y., Dong, S. Y., He, Q., Jia, Y., Zhao, L. N., Yu, S. X., Liu, H. Y., Liu, B., Yan, Y. H., Xiang, J. Y., Xia, N. H., Peng, H., Li, Z. Y., Zhang, Z. X., He, X. J., Yin, L. K., Lin, Y. L., Liu, Q. R., . . ., Xue, N. X. (2017). Threatened species list of China's higher plants. *Biodiversity Science, 25*, 696–744. (In Chinese).

Quan, G. Q., & Xie, J. H. (2002). *Research on the golden monkey*. Shanghai Science and Technology Education Press (In Chinese).

Salinas, N., Malhi, Y., Meir, P., Silman, M., Roman Cuesta, R., Huaman, J., Salinas, D., Huaman, V., Gibaja, A., Mamani, M., & Farfan, F. (2011). The sensitivity of tropical leaf litter decomposition to temperature: Results from a large-scale leaf translocation experiment along an elevation gradient in Peruvian forests. *New Phytologist, 189*, 967–977.

Sandberger-Loua, L., Doumbia, J., & Rödel, M. O. (2016). Conserving the unique to save the diverse-identifying key environmental determinants for the persistence of the viviparous Nimba toad in a West African World Heritage Site. *Biological Conservation, 198*, 15–21.

Sanders, N. J., Lessard, J. P., Fitzpatrick, M. C., & Dunn, R. R. (2007). Temperature, but not productivity or geometry, predicts elevational diversity gradients in ants across spatial grains. *Global Ecology and Biogeography, 16*, 640–649.

Schnell, J. K., Harris, G. M., Pimm, S. L., & Russell, G. J. (2013). Estimating extinction risk with metapopulation models of large-scale fragmentation. *Conservation Biology, 27*, 520–530.

Shan, Z. J., Zhao, L. N., Yang, Y. C., Xie, D., & Qin, H. N. (2019). An overview on assessment systems for threatened plants in China. *Biodiversity Science, 27*, 1352–1363. (In Chinese).

Shen, G. Z., Feng, C. Y., Xie, Z. Q., Ouyang, Z. Y., Li, J. Q., & Pascal, M. (2008). Proposed conservation landscape for giant pandas in the Minshan Mountains, China. *Conservation Biology, 22*, 1144–1153.

Silveira, L., Sollmann, R., Jácomo, A. T., Diniz-Filho, J. A., & Tôrres, N. M. (2014). The potential for large-scale wildlife corridors between protected areas in Brazil using the jaguar as a model species. *Landscape Ecology, 29*, 1213–1223.

Sundqvist, M. K., Sanders, N. J., & Wardle, D. A. (2013). Community and ecosystem responses to elevational gradients: Processes, mechanisms, and insights for global change. *Annual Review of Ecology, Evolution and Systematics, 44*, 261–280.

The Operational Guidelines for the Implementation of the World Heritage Convention. (n.d.). http://whc.unesco.org/en/guidelines/

Thorsell, J. (2009). A global overview of mountain protected areas on the world heritage list. *Mountain Research and Development, 34*, 291–291.

Tie, J. (2013). *The plant composition in the habitat and food resources of Rhinopithecus roxellana in Shennongjia, China*. China Forestry Publishing House (In Chinese).

Trent, J. A. (2010). *Ecology, habitat use, and conservation of Asiatic black bears in the Min mountains of Sichuan province, China* [Master dissertation]. Virginia Polytechnic Institute and State University.

Udvardy, M. D. (1975). *A classification of the biogeographical provinces of the world*. INCN Paper No. 18. Morges, Switzerland, pp. 20–21. Retrieved December 19, 2016, from http://agris.fao.org/agris-search/search.do?recordID=XF2016014050.

Wagai, R., Kitayama, K., Satomura, T., Fujinuma, R., & Balser, T. (2011). Interactive influences of climate and parent material on soil microbial community structure in Bornean tropical forest ecosystems. *Ecological Research, 26*, 627–636.

Wang, C. L., Zang, Z. H., Qiu, Y., Deng, S. Y., Feng, C. Y., Xie, Z. Q., Xu, W. T., Liu, L., Chen, Q. S., & Shen, G. Z. (2017). The effectiveness of Shennongjia national nature reserve in conserving forests and habitat of Sichuan snub-nosed monkey. *Biodiversity Science, 25*(5), 504–512. (In Chinese).

Wang, Y. I., Li, X. L., & Liu, C. Q. (2014). A prove regarding Sanqingshan. In *World Natural Heritage sustainable development*. Chinese Garden, pp. 85–87.

Wang, H., Lü, Z., Gu, L., & Wen, C. (2015). Observations of China's forest change (2000–2013) based on global forest watch dataset. *Biodiversity Science, 23*, 575–582. (In Chinese).

Wardle, D. A. (2002). Communities and ecosystems: Linking the aboveground and belowground components. Princeton: Princeton University Press.

Wu, N., Lei, B. Y., Zhou, Y. B., & Xie, Z. Q. (2019). *The catalogue of animal type specimens in Shennongjia, Hubei Province, China*. Science Data Bank (In Chinese).

Wu, Z. Y. (2010) .*Chinese floristic geography*. Science Press (In Chinese).

Xiang, Z. F., Yu, Y., Yang, M., Yang, J. Y., Liao, M. Y., & Li, M. (2011). Does flagship species tourism benefit conservation? A case study of the golden snub-nosed monkey in Shennongjia national nature reserve. *Chinese Science Bulletin, 56*, 2553–2558.

Xie, Z. Q., & Shen, G. Z. (2018). *The value of Shennongjia natural heritage and its management*. Science Press (In Chinese)

Xie, Z. Q., Shen, G. Z., Zhou, Y. B., Fan, D. Y., Xu, W. T., Gao, X. M., Du, Y. J., Xiong, G. M., Zhao, C. M., Zhu, Y., & Lai, J. S. (2017). The outstanding universal value and conservation of the Shennongjia world natural heritage site. *Biodiversity Science, 25*(5), 490–497. (In Chinese).

Xie, Z. Q., & Xiong, G. M. (2020). *The catalogue of type specimen plants in Shennongjia*. Science Press (In Chinese).

Xie, Z. Q., Xu, W. T., Shen, G. Z., & Xiong, G. M. (2020). Long term monitoring to promote biodiversity conservation and ecological construction in subtropical mountainous areas. *Bulletin of Chinese Academy of Science, 35*(9), 1189–1196. (In Chinese).

Xiong, G. M., Shen, G. Z., Fan, D. Y., Xu, W. T., Zhou, Y. B., Zhao, C. M., & Xie, Z. Q. (2017). Community participation in Hubei Shennongjia natural heritage-current status and countermeasures. *Shanaxi Forestry Science and Technology, 5*, 26–31. (In Chinese).

Xu, W.T., Ge, J. L., Xiong, G. M., Zhao, C. M., Shen, G. Z., Fan, D. Y., Zhang, M., Liu, J. Y., & Xie, Z. Q. (2017). A *dataset of species composition in a typical subtropical mixed evergreen and deciduous broad-leaved forest*. Science Data Bank (In Chinese).

Xu, W. H., Ouyang, Z. Y., Viña, A., Zhen, H., Liu, J. G., & Xiao, Y. (2006). Designing a conservation plan for protecting the habitat for giant pandas in the Qionglai mountain range, China. *Diversity and Distributions, 12*, 610–619.

Xu, W. T., Xie, Z. Q., Shen, G. Z., Zhou, Y. B., Zhao, C. M., Xiong, G. M., & Ge, J. L. (2019). The nature and geography scope of Shennongjia. *Territory & Natural Resources Study, 3*, 42–46. (In Chinese).

Yan, L., Xu, X. G., & Zhang, X. P. (2009). Analysis to temporal characteristics of tourist flows on Jiuzhaigou World Natural Heritage. *Acta Scientiarum Naturalium Universitatis Pekinensis, 45*(1), 171–177.

Yang, M. X. (2008). Analysis of Zhangjiajie tourism income and its enlightenment. *Social Scientist, 26*(6), 99–103.

Yang, G. R., & Xie, C. X. (1983). Preliminary study on fish in Shennongjia. *Chinese Journal of Zoology, 6*, 35–36. (In Chinese).

Ying, J. S. (2001). Species diversity and distribution pattern of seed plants in China. *Biodiversity Science, 9*, 393–398. (In Chinese).

Ying, J. S., Ma, C. G., & Zhang, Z. S. (1979). Vegetation and flora of Shennongjia region in western Hubei. *Acta Phytotaxonomica Sinica, 3*, 41–60. (In Chinese).

Ying, K. P. (2010). *Tracing on hundred years of change: Illustrating the environment changes in western China*. Encyclopedia of China Publishing House (In Chinese).

Yu, J., Wu, N., & Zhou, Y. B. (2019). *The birds of Shennongjia*. Science Press (In Chinese)

Yu, X. L., Han, W. B., Zhou, Y. B., Shen, G. Z., Xu, W. T., & Xie, Z. Q. (2018a). Technical parameters for designing crossings for the Sichuan snub–nosed monkey (Rhinopithecus roxellana) in Shennongjia world natural heritage site. *Ecological Science, 37*(4), 97–104. (In Chinese).

Yu, X. L., Zhou, Y. B., Shen, G. Z., Xiong, G. M., Xu, W. T., & Xie, Z. Q. (2018b). Tourism environmental carrying capacity of Shennongjia world natural heritage site. *Ecological Science, 37*(1), 158–163. (In Chinese).

Yu, X. L., Zhou, Y. B., Xu, W. T., & Xie, Z. Q. (2015). Principle and technical parameters for designing wildlife crossings on tourism roads in protected areas. *Biodiversity Science, 23*, 824–829. (In Chinese).

Zhang, Y. B., & Ma, K. P. (2008). Geographical distribution patterns and status assessment of threatened plants in China. *Biodiversity and Conservation, 17*, 1783–1798.

Zhang, Y. Y., Zhang, Z. W., Dong, L., Ding, P., Ma, Z. J., & Zheng, G. M. (2016). Assessment of red list of birds in China. *Biodiversity Science, 24*, 568–577. (In Chinese).

Zhao, C. M., Han, W. B., Shen, G. Z., Fan, D. Y., Xiong, G. M., Zhou, Y. B., Xu, W. T., & Xie, Z. Q. (2018). The impact of the opening of Shennongjia airport and the listing of world heritage on tourism increase. *Ecological Science, 37*(3), 138–142. (In Chinese).

Zhao, C. M., Chen, W. L., Tian, Z. Q., & Xie, Z. Q. (2005). Altitudinal pattern of plant species diversity in Shennongjia mountains, central China. *Journal of Integrative Plant Biology, 47*, 1431–1449.

Zheng, D. L., & Yuan, J. H. (2009) SWOT analysis and development suggestions of Taishan tourism industry. *Journal of Shandong Institute of Business and Technology, 23*(1), 34–38.

Zhou, Q. J., Han, W. B., Zhao, C. M., Wang, Y., Chen, F. Q., & Xie, Z. Q. (2019). Species composition and community structure of natural mixed needle- and broad-leaved forest in Shennongjia. *Chinese Journal of Ecology, 38*(1), 11–18. (In Chinese).

Zhou, Y. B., Han, W. B., Chen, W. W., Cui, J. F., Shen, G. Z., Xiong, G. M., Xu, W. T., Fan, D. Y., Zhao, C. M., & Xie, Z. Q. (2018). Terrestrial vertebrate diversity in Shennongjia world natural heritage site, China. *Ecological Science, 37*(5), 47–52. (In Chinese).

Zhou, Y. B., & Lei, B. Y. (2019). *The vertebrate of Shennongjia*. Science Press (In Chinese).

Zhou, Y. B., & Wu, N. (2019). *The catalogue of animal type specimens in Shennongjia*. Science Press (In Chinese).

Zhou, Y. B., Xu, W. T., Zhao, C. M., Shen, G. Z., Xiong, G. M., Fan, D. Y., & Xie, Z. Q. (2017a). Feasibility and technical design for enhancing habitat connectivity across two separated components in Shennongjia world natural heritage site, China. *Chinese Journal of Ecology, 36*(10), 2988–2996. (In Chinese).

Zhou, Y. B., Yu, X. L., Wu, N., Shen, G. Z., Xiong, G. M., Xu, W. T., Fan, D. Y., Zhao, C. M., & Xie, Z. Q. (2017b). A catalogue of animal type specimens from the Shennongjia world natural heritage site, China. *Biodiversity Science, 25*(5), 513–517. (In Chinese).

Zhu, Q., Yu, K. J., & Li, D. H. (2005). The width of ecological corridor in landscape planning. *Acta Ecologica Sinica, 25*, 2406–2412. (In Chinese).

Zhu, Z. Q., & Song, C. S. (1999). *Scientific survey of Shennongjia natural reserve*. Beijing Forestry Press (In Chinese).

The manufacturer's authorised representative in the EU is Springer Nature Customer Service Centre GmbH, Europaplatz 3, 69115 Heidelberg, Germany. If you have any concerns regarding our products, please contact ProductSafety@springernature.com

Printed and bound by CPI Group (UK) Ltd, Croydon, CR0 4YY
30/09/2026
02225294-0001